T0302013

THE KYOTO 京 都 MODEL

Akira Ishikawa
Aoyama Gakuin University, Japan

Koji Tanaka
Mitsubishi Logistics, Japan

THE KYOTO 京 都 MODEL

The Challenge of Japanese Management Strategy Meeting Global Standards

 World Scientific

NEW JERSEY · LONDON · SINGAPORE · BEIJING · SHANGHAI · HONG KONG · TAIPEI · CHENNAI

Published by

World Scientific Publishing Co. Pte. Ltd.

5 Toh Tuck Link, Singapore 596224

USA office: 27 Warren Street, Suite 401-402, Hackensack, NJ 07601

UK office: 57 Shelton Street, Covent Garden, London WC2H 9HE

Library of Congress Cataloging-in-Publication Data
Ishikawa, Akira, 1934–
    [Kyoto moderu. English]
    The Kyoto model : the challenge of Japanese management strategy meeting global
standards / by Akira Ishikawa and Koji Tanaka.
        p. cm.
    Includes bibliographical references and index.
    ISBN-13 978-981-256-329-3 (alk. paper)
    ISBN-10 981-256-329-6 (alk. paper)
        1. High technology industries--Japan--Kyoto--Management. 2. Industrial
management--Japan--Kyoto. I. Tanaka, Koji, 1967–   II. Title.

HC463.K9I8413 2005
658.4'00952'186--dc22

                                                                                2005041832

British Library Cataloguing-in-Publication Data
A catalogue record for this book is available from the British Library.

Cover graphics by Noriko Nakagawa / Office Square, Japan
Translation by Katsumi Matsubara and Ulrich Manz / TranNet KK, Japan
Typeset by Stallion Press / Email: enquiries@stallionpress.com

Printed in Singapore

Preface

This book originated from discussions on the future course of Japanese companies and the limitations of Japanese business management style in a class on international accounting for the Master of Business Administration (MBA) course of Aoyama Gakuin University. Discussions focused on the perception that Japanese companies and the Japanese style of business management may no longer be effective in the ever-changing global market.

Some people there noted that while some views focused on the shortcomings of Japanese companies, not an insignificant number of Japanese companies have captured leading global market shares. We then conducted some research and found that a considerable number of Japanese companies have obtained leading market shares with globally preeminent technologies and management systems in areas that we usually overlook.

As we looked into such facts, we began to question the conclusion that "Japanese companies or Japanese-style management would no longer work." We began to believe that Japanese manufacturers boasting the world's leading technologies and management systems may hold the key to the revival of the Japanese economy.

Among such excellent Japanese companies, this book focuses on Kyoto high-tech companies that have captured leading market shares in their respective areas and maintained their robust profitability even under the current prolonged recession.

Needless to say, Kyoto had been an administration center for more than 1,000 years in Japan's history. In the ancient Japanese capital, many unique Japanese cultures flourished. Our research targets included Kyocera Corp., Omron Corp., Murata Manufacturing Co., Rohm Co., and Horiba Ltd., which were founded in Kyoto and have fast developed into global companies.

Why have these Kyoto high-tech companies achieved high growth and maintained their growth even amid the current recession? What are their secrets? This book's objective is to give consideration to a new Japanese business management model through an analysis of these companies' management systems.

We have decided to generally call the Kyoto high-tech companies' management systems the "Kyoto Model."

The Kyoto Model means management systems where these companies have developed market-oriented products or technologies efficiently and speedily under the strong leadership of their founders and their successors. They have successfully produced innovations by boosting their employees' morale. Supported by their excellent production technologies, these companies have continued to develop competitive products. These firms have also positively promoted the disclosure of financial information as a result of the priority they give to the stock market and stepped up various philanthropic activities while pursuing the sustainable growth of themselves and society as a whole. These features may be those of a model that has the potential to be accepted globally.

The Kyoto high-tech companies have continued to increase revenues and profit even under the current recession while maintaining top Japanese and global market shares in their respective core areas. This is because the closed Japanese market has prompted these companies to pursue openness and explore overseas (especially the U.S.) markets since their early days for both their survival and growth.

These high-tech companies have developed their present management systems while reforming their systems flexibly for their adaptation to overseas markets and their survival in the Japanese market. The fruit of their continuous efforts is the Kyoto Model.

The U.S. business management system has emphasized bold strategic shifts under top managers' strong leadership, improvement of productivity, and efficient exploitation of capital resources, while the Japanese management system has given priority to continuous improvement of productivity and production technologies, and personnel development in terms of value added under long-term perspectives. The Kyoto Model is a unique management system combining these strengths of the U.S. and Japanese management systems.

The Kyoto Model shares many points in common with the management systems of the so-called visionary U.S. companies including 3M and Hewlett-Packard. These U.S. firms, renowned for their introduction of Japanese management approaches, have developed into global companies and are over 60 years old (3M is nearly 100 years old), but they still have the features of venture businesses.

The Kyoto high-tech companies are neither as old nor as global as 3M and Hewlett-Packard, but their potential to grow further can be seen in their management systems.

R&D-oriented high-tech ventures are expected to play a key role in reviving and enhancing the Japanese economy in the 21st century. The Kyoto Model is a business management model for Japan's R&D-oriented ventures. In this sense, the Kyoto Model as a case study subject is expected to become a useful tool for the management of the R&D-oriented high-tech ventures that will support the Japanese economy in the 21st century.

The Introduction analyzes what Japanese-style business management is. By comparing the Japanese and U.S. styles of business management, this chapter is designed to specify the Japanese style's strengths that Japanese companies should retain as well as the U.S. style's strengths that they should adopt.

Chapter 1 reaffirms the excellent performance of the Kyoto high-tech companies based on their financial data and outlines the Kyoto Model.

Chapter 2 discusses the reasons why these high-tech ventures were founded in the ancient Japanese capital of Kyoto and have developed into global companies capturing top market shares.

The subsequent chapters detail the Kyoto Model. Chapter 3 deals with the Kyoto high-tech companies' history, beginning with their founding, Chapter 4 with their management attitude, giving priority to corporate philosophies, Chapter 5 with their organization management, and Chapter 6 with the emphasis they put on R&D management and production technologies, and their personnel management. Finally, Chapter 7 focuses on their philanthropic activities.

The Kyoto high-tech companies have established their respective unique management styles under top managers, who have demonstrated their strong leadership. Some may doubt the advisability of combining these different management styles into a single business management model. The reason why we have combined these companies into the Kyoto Model is that they are the first of the R&D-oriented high-tech ventures expected to support the Japanese economy in the future. They feature the efficient maximization of business resources to select and focus on core products to be successful in global markets, and have many key points that other Japanese firms in general should learn from.

But this book has focused on only a small number of the companies founded in Kyoto. There are many other unique Kyoto-based companies that have also been successful and have won leading market shares in their respective areas. These and the Kyoto Model companies may have some features in common.

In future, we may have to look at not only Kyoto but also other regional economies in Japan. We suspect that Fukuoka, Yokohama, and other cities may have companies faring well in areas we usually overlook. Case studies on such companies may provide key hints of a new Japanese management style.

We feel that many Japanese companies are losing their identity amid the drastically changing business environment.

Writer Ryotaro Shiba says of Japanese people's adaptation to social changes:

> *Since the Taika Reform (in 645), the Japanese have repeatedly reformed their society. The Japanese are one of a relatively few races that have done so. The Japanese are capable of doing so. The power of Japanese culture stems from that capability. (Hand-Cut Japanese History, p. 128)*

In response to the business environment changes currently facing Japanese companies, they may be able to reform their Japanese-style business management to recover their strength.

We hope that this book will help reinvigorate Japanese companies that are losing their strength.

Last, we thank a number of people for kindly helping us publish this book. Kyocera Corp., Ms. Akiko Yamada from the Personnel Management Group of Omron Corp., the Public Relations Office of Murata Manufacturing Co., Mr. Yasushi Suzuki from the Public Relations Office of Rohm Corp., Mr. Tetsuhiro Habu from the Public Relations Office of Horiba Ltd., and Study Group I have kindly responded to our questionnaire surveys, sent data to us, and given constructive comments.

We also heartily thank Mr. Yoshiaki Toya of Prentice Hall for giving us enormous support and assistance from the beginning of our book-writing process.

Akira Ishikawa
Koji Tanaka
March 1999

About the Authors

Akira Ishikawa

Professor Emeritus and former Dean, Graduate School of International Politics, Economics and Business, Aoyama Gakuin University. Studied at the University of Washington and University of Texas, Graduate School of Business Administration, with postdoctoral studies at Massachusetts Institute of Technology. Served as a lecturer at University of Texas, an assistant professor at the Graduate School of New York University, and a professor at the Graduate School of Rutgers University. Authored *Introduction to Strategic Information Systems, Strategic Budget Management, Why do Conferences Fail?*, and others. Coauthored *Modern International Accounting, The Success of 7-Eleven Japan* and *Top Global Companies in Japan*. Translated *Managing Chaos, Defense Management*, and others.

Koji Tanaka

Graduated from Economics Faculty, Keio University. Completed an M.A. in international politics and economics at Aoyama Gakuin University (Master of International Business Administration). Working at Mitsubishi Logistics Corp. Translated *Logistics and Supply Chain Management*.

Contents

Preface   v

About the Authors   xi

**Introduction: Japanese and U.S. Styles of
Business Management**   1

Features of Japanese-Style Management   1
U.S.-Style Management   4
Strengths and Weaknesses of Japanese and U.S. Business
    Management Styles   5
Kyoto High-Tech Companies   11

**Chapter 1 The Kyoto Model: Kyoto High-Tech
Companies' Management Systems**   13

Sales and Profits Increase Even Amid Recession   13
Top Leadership   18
Organization and Control Systems   21
Philanthropy   25

**Chapter 2 Why Were the High-Tech Companies
Founded in Kyoto? — Naturally or Accidentally?**   27

Analysis Based on Macro Data   27
Kyotoism   31
Sustainable Business Management   32
Absence of Big Banks and Companies   34

**Chapter 3 Real Faces of Kyoto High-Tech Companies:
History of Growth Since Foundation**   37

Core-Based Growth   37
Kyocera: From Ceramic Components Manufacturer to
    Integrated Circuits Manufacturer   37
Omron: Grasping Social Needs   40
Murata: Focusing on Ceramics   41
Rohm: Bold Challenging Approach on ICs   42
Horiba: Originating from a Student Venture   44
Keeping Top Market Shares for Core Products   45
Aggressive Globalization   47
Expansion through M&A Deals   50
Aggressive, Bold Acquisitions — Kyocera   50
Enhancing Information Technologies — Omron   52
Upgrading Core Competences — Murata, Rohm,
    Horiba   53

**Chapter 4 Corporate Philosophies Emphasized
in Management**   55

Kyoto High-Tech Companies Emphasize
    Corporate Philosophies   55
How Have Corporate Philosophies Been Developed? . .   58
Communication of Corporate Philosophy   62

Chapter 5 Unique Organizational Management   65

Exploitation of Information Technology   66
Kyocera: Ameba Management   68
Omron: From Producer System to Holding
    Company   71
Murata: Matrix Management of Smaller Units to
    Control   76
Rohm: Horizontal Organization of Small Units   78
Horiba: Product-Oriented Management System   80

Chapter 6 R&D and Production Control, Cost Control and Personnel Management Systems 83

Introduction 83
Market-Oriented R&D 83
Focusing on Production Technologies 86
Maximizing Employees' Value Added 88
Exercising Innovation 89
Incentives for Learning 93

Chapter 7 Positive Social Contributions 95

Accountability 95
Philanthropy 96
Tackling Global Environmental Problems 98
New Venture Development 100

Chapter 8 Conclusion 103

Global Companies Born in the Ancient Capital of
Kyoto 103
Pressure of Slowing Economic Growth 106
Learning from U.S.-Style Business Management
Again 107

Appendix 109

Bibliography 115

Index 119

Japanese and U.S. Styles of Business Management

Features of Japanese-Style Management

Japanese-style business management has demonstrated its strength in processing and assembly industries including electrical machinery and automobiles. Japanese companies have had a competitive edge in production technologies and systems. Their competitiveness has stemmed from their production systems, which have efficiently turned out high-quality products.

Japanese-style management attracted world-wide attention in the early 1980s. As the sluggish performance of U.S. companies prompted U.S. managers to review their management style and look for measures to revive U.S. manufacturing companies, Japanese-style management became the subject of massive research.

It may be difficult to define Japanese-style management simply. Generally, lifetime employment, seniority-based salaries and promotions, and in-house unions are frequently cited as features of Japanese-style management.

Prof. Tadao Kagono describes Japanese-style management as follows:

Japanese-style management means the business management concept or system based on Japanese culture and institutional conditions. The foundation of Japanese-style management includes employment systems like lifetime employment, seniority-based salaries and promotions, and in-house unions, long-lasting transactions between companies, and pluralistic corporate governance mainly by insiders. Based on these systems, Japanese companies have created unique business management systems including the ethos of management from the long-term viewpoint, quality control, the kanban just-in-time system and concurrent development of products. (Recovery of Japanese-Style Management, p. 2.)

We must pay attention to the fact that lifetime employment does not mean employment of people over their entire lifetimes. It means that companies may not fire employees unilaterally before they reach the mandatory retirement age, unless the companies face a financial crisis. Under a recession like the present one, Japanese companies are effectively dismissing employees before they reach retirement age limits, using early retirement and other systems for corporate restructuring. This is a fact that is more conspicuous among small- and medium-sized enterprises. On the other hand, some large European and U.S. companies' employment systems for white-collar workers are reportedly close to Japan's lifetime employment. In this sense, lifetime employment may not necessarily be unique to Japan.

The reason why lifetime employment is often cited as the main feature of Japanese-style management, despite these facts, may be the attention focused on the various benefits that result from a combination of lifetime employment and seniority-based salaries and promotions. Long-term employment allows employee job rotation, which can trigger sophisticated sharing of information and accumulation of knowledge in all business activities ranging from production to research and development, marketing and accounting. This information sharing and knowledge accumulation has played a great role in giving Japanese companies a competitive edge.

Here is a concept that employees and their knowledge are a kind of capital stock. This means that companies, only by retaining employees over the long term, can take full advantage of their employees' productivity to keep their management and production efficient. In this sense, we may be able to conclude that lifetime employment is a very reasonable system to achieve optimum exploitation of human resources. (Odagiri, *Japanese Companies' Strategy and Organization*, p. 5.)

For Japanese companies, especially globally successful shipbuilders, automakers, and electrical machinery builders in the processing and assembly industries, the knowledge and know-how accumulated among production plant workers has played a significant role in their growth. Knowledge and know-how accumulated in an organization can deepen through communication and coordination among employees and bring about innovations in the organization.

Japanese companies have taken advantage of the knowledge developed by their employees for products, services, or business systems on a company-wide basis. The organized knowledge development has been the largest contributor to the success of Japanese companies and the key to Japanese-style innovations. (Nonaka and Takeuchi, *Knowledge-Developing Companies*, p. 1.)

The continuous accumulation and conversion of knowledge, as illustrated in Figure 1, is the foundation of Japanese companies' competitive advantage, allowing them to continuously innovate and grow.

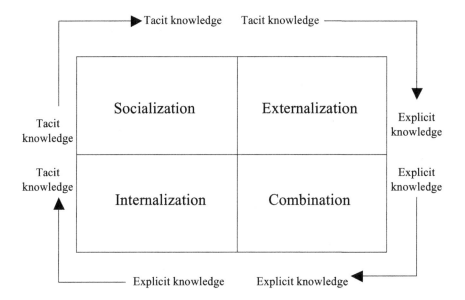

Fig. 1.   Four knowledge conversion modes (From *Knowledge-Developing Companies*, p. 93)

U.S.-Style Management

Next, we would like to consider U.S.-style management.

American business managers give top priority to maximizing returns on efficient investment of capital borrowed from shareholders. Managers may be dismissed by shareholders if they fail to achieve sufficient returns. U.S. business managers are always put under heavy pressures from shareholders.

U.S. companies typically do not have cross-shareholding, main bank, and other systems that Japanese companies possess to defend themselves from takeovers. As far as their equity shares are publicly traded, they are always exposed to the risk of takeover. Their managers have no choice but to keep share prices high in order to defend their companies from takeovers.

Faced by shareholders' demands for better profitability and the capital market's pressures to keep stock prices high to deter takeovers, U.S. business managers seek to obtain higher returns in short periods of time and consequently make short-term management decisions. The capital market's pressures prompt U.S. managers to use layoffs and other restructuring measures in the face of a slowdown in earnings to recover profitability in a short period of time.

U.S. workers for their part pursue jobs to take advantage of their expertise, rather than be willing to belong to companies as organizations. They are willing to be professionals backed by their expertise, rather than be loyal to companies. U.S. workers thus easily change companies. Companies employ expertise-owning workers when they are necessary. As a whole, the United States has a high level of job mobility.

Such a background allows U.S. companies to drastically change business strategies through their top-down decision-making. Through the replacement of managers, restructuring, and other measures, strategies can be changed drastically.

Changes in business strategies affect the products being produced and services being provided. A typical U.S. company aims at developing a new product meeting its strategy and at obtaining a top share of a new market with the product. Under U.S.-style management, epoch-making products that are devised free from existing ones can be developed discontinuously.

Based on such U.S.-style management, a typical U.S. company has a hierarchic organization where top-down communication can be efficiently enacted. Such a hierarchic organization is easy to reform and is suitable for top managers' strategy shifts.

The greater job mobility has led U.S. companies to develop advanced business manuals for knowledge communication. This is far different from Japanese companies' organized communication and accumulation of knowledge as "tacit knowledge."

Strengths and Weaknesses of Japanese and U.S. Business Management Styles

Regarding the differences between the Japanese and U.S. styles of business management, Prof. Tadao Kagono and his group focus on how companies adapt themselves to the business environment. They compare the Japanese operation-oriented strategy with the U.S. product-oriented one, and the Japanese group dynamics with the U.S. bureaucratic dynamics (Kagono *et al.*, *Comparison between Japanese and U.S. Business Management Styles*).

The operation-oriented strategy accumulates knowledge while giving priority to production strategies and is very adaptable to continuous business environment changes. On the other hand, the product-oriented strategy develops business resources flexibly while giving priority to product strategies and is very adaptable to rapid business environment changes (see Table 1).

Group dynamics is an organizing approach to achieve organizational integration through frequent interaction between individual members and groups based on their shared values and information within an organization. Bureaucratic dynamics is an approach to build a formalized organizational hierarchy and attain organizational integration through rules and plans (see Table 2).

Table 1. Operation-oriented strategy and product-oriented strategy

	Operation-oriented strategy	Product-oriented strategy
Strength	Adaptability to continuous changes (exact adaptation to changes) A variety of strategy elements can be precisely combined	Resources can be distributed efficiently Adaptability to drastic structural changes A major breakthrough can be produced
Weakness	Distribution of resources could be inefficient Less adaptability to drastic structural changes A major breakthrough is difficult to produce	Less precise adaptability to continuous changes Companies may fall into complacent strategies due to their excessive pursuit of uniqueness in product concepts

As indicated by the above, the Japanese style of business management gives priority to the group dynamics approach to organization based on the operation-oriented strategy. In contrast, the U.S. style emphasizes the bureaucratic dynamics approach based on the product-oriented strategy.

We must, however, pay attention to the fact that all companies in Japan and the United States adapt themselves differently to the business environment. Even within each country, companies differ in strategic orientation and organization.

Firms that adopt the operation-oriented strategy are good at developing each product into a major one through continuous improvements based on accumulated knowledge. This strength has supported electrical machinery, automobile and other processing and assembly sectors as Japan's industrial leaders. With the operation-oriented strategy, however, companies are not good at making bold strategic shifts to develop innovative products.

Japanese companies that emphasize group dynamics as an organizing approach focus on fine adjustments to meet continuous changes. This is their strength. On the other hand, their weakness

Table 2. Group dynamics and bureaucratic dynamics

	Group dynamics	Bureaucratic dynamics
Strength	Organization units can autonomously adapt to environmental changes to prevent information being concentrated at the top management level Organization units are encouraged to practice learning to promote accumulation of knowledge and experience in the workplace Values of an organization can be used to motivate employees and draw on their great psychological energy	Hierarchization of an organization can be promoted to cope with great information-processing loads Knowledge can be accumulated within an organization irrespective of personnel replacements Top managers' strategic decisions to restructure an organization can reform employees' action patterns within a short period of time
Weakness	As information-processing and communication channels are uncertain, communication shortages and other organizational confusion can emerge As knowledge accumulation is made on an individual basis, an organization is vulnerable to personnel replacements It takes a lot of time to change employees' action patterns	Hierarchization of an organization can impose constraints on lower-level employees' freedom As horizontal coordination between divisions is difficult, information-processing loads concentrate on top managers

results from their lack of the kind of drastic strategic development capabilities possessed by U.S. firms that give priority to the bureaucratic dynamics approach.

Given the above, the challenge for Japanese-style management is to increase their adaptability to drastic business environment changes. Specific measures to this end may include the introduction of elements of the product-oriented strategy and the bureaucratic dynamics approach.

It must be remembered that the strength of Japanese-style management has stemmed from an organized knowledge-development process. Therefore, what Japanese companies should do for their improvement may be to introduce elements of the product-oriented strategy.

Table 3 illustrates these characteristics, comparing and categorizing leading Japanese and U.S. companies. In accordance with environmental adaptation patterns, companies are divided into four categories — horizontal adaptation, vertical adaptation, bureaucratic adaptation, and symbiotic adaptation. The vertical adaptation should be highlighted. As indicated in Figure 2, companies close to the vertical adaptation pattern include 3M and Hewlett-Packard in the United States and Kyocera and TDK in Japan.

A vertical-adaptation company is an organization of small teams, with its top manager demonstrating strong entrepreneurial leadership. Each team is highly autonomous. Knowledge and know-how accumulation depends on individuals' or teams' learning. Knowledge and know-how are inherited as success stories within the organization and imitated.

A vertical-adaptation company makes innovations experimentally and views risk as a positive challenge. The vertical adaptation is suitable for a risky environment where drastic and unpredictable changes emerge continuously. A typical vertical-adaptation company is a high-tech venture.

In the present business environment where unpredictable changes are coming one after another, Japanese companies may have to make use of such vertical adaptation. Vertical adaptation includes the strength of the product-oriented strategy's adaptation to drastic environment changes while taking advantage of the strength of the group dynamics approach.

As noted earlier, the challenge facing Japanese companies is to tackle the product-oriented strategy while taking advantage of the strength of the group dynamics approach. The problem for Japanese-style management now may be how to introduce elements of vertical adaptation.

Table 3. Features of four environmental adaptation patterns

	Group dynamics	Bureaucratic dynamics	
Operation-oriented strategy	Horizontal adaptation (1) Frequent interaction, sharing of values and information, creation of tensions, personal communications networks (2) Loose organization, in-house power decentralization (3) Interactive learning and sharing of learning within an organization (4) Missionary leadership (promoting sharing of directions and philosophies) (5) Passive, inductive, and incremental adaptation (6) Operational efficiency, minimized product characteristic gaps, synergy, speedier adaptation (7) Workplace information, information through contacts with customers (8) Sense of unity, harmony	Bureaucratic adaptation (1) Rules, programs, hierarchy, functional division of labor, job-based salaries (2) Functional organization, concentration of power in top managers and elite staff (3) Elite-based learning, systems, manuals, rule-based communications (4) Technocrat leadership (5) Passive and quantitative adaptation, defense of domains (6) Production efficiency, cost advantage (7) Quantitative information (8) Compliance with rules, duties, and procedures	(1) Organizational integration and information-processing means (2) Distribution of influences and organizational forms (3) Accumulation and conveyance of knowledge and information (4) Top Management leadership (5) Methods to cope with opportunities and threats (6) Keys to environmental adaptation and competitive advantage (7) Information pattern (8) Value pattern
Product-oriented strategy	Vertical adaptation (1) Frequent interaction, sharing of value information, teams, task forces, commitments to technologies and products (2) Organization of small teams, dispersion of decision-making centers (3) Team or individual learning, sharing of learning through modeling (4) Entrepreneurial leadership (5) Active, experimental, and continual adaptation (6) Uniqueness of products, leading to innovation (7) Fresh customers, information technology (8) Challenge to risks, venture spirit	Symbiotic adaptation (1) Hierarchy, self-contained organization, vertical information channels, plans, targets, performance-based salaries (2) Divisional organization, concentration of resource distribution authority in top management (3) Elite-based learning, system-based information conveyance, acquisition of businesses (4) Shogun leadership (5) Deductive, analytical and systematic adaptation (6) Strategic persistence (7) Systematic predictions (8) Persistence, compliance with plans, achievement of targets	

(From *Comparison between Japanese and U.S. Business Management Styles*, p. 233)

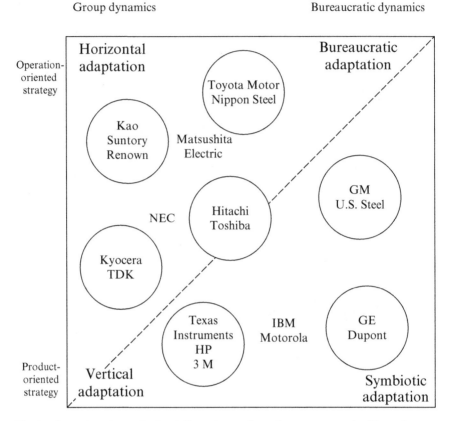

Fig. 2.   Strategic and organizational dimensions and gaps between companies (From *Comparison between Japanese and U.S. Business Management Styles*, p. 229)

The elements of vertical adaptation may weaken as a company grows larger with products more diversified. Therefore, large companies have difficulties in maintaining vertical adaptation. A vertical-adaptation company has higher growth potential, but it may have difficulties in maintaining vertical adaptation elements as it grows larger. Such a dilemma is a major management issue that always exists in any vertical-adaptation company.

In this regard, large vertical-adaptation companies adopt the top managers' strong personality or employees' sharing of the company mission, the establishment of smaller business units, and other measures to overcome the constraints accompanying their growth. In order to maintain vertical adaptation, companies must make continuous management-improving efforts that may consume enormous amounts of labor and time.

Kyoto High-Tech Companies

Apart from Kyocera, which is categorized as close to the vertical-adaptation model along with TDK, 3M, and Hewlett-Packard, other Kyoto high-tech companies including Omron, Murata, Rohm, and Horiba have many vertical-adaptation elements.

In these Kyoto high-tech companies, the founders and other top managers have demonstrated strong entrepreneurial leadership. Each of these companies consists of small teams that are highly autonomous. Accumulation of knowledge and know-how depends on individuals' and teams' learning based on vigorous in-house communication.

In order to address a decline in vertical adaptation elements emerging with their growth in size and product diversification, the Kyoto high-tech companies have had corporate philosophies shared among employees under the leadership of their top managers and have minimized the size of units to control.

Next, we would like to analyze the Kyoto Model as the management system of the Kyoto high-tech companies.

Chapter 1

The Kyoto Model: Kyoto High-Tech Companies' Management Systems

Sales and Profits Increase Even Amid Recession

First, we would like to look at the business performance of the Kyoto high-tech companies as indicated in their earnings reports. In fact, the Kyoto high-tech companies have been increasing both sales and profits even amid the current recession.

As shown in Figure 3, consolidated sales of the Kyoto high-tech companies have continued to increase over the past decade, excluding the demand-shrinking period just after the burst of the economic bubble and the 1993–1995 turbulence caused by the yen's rapid appreciation. From fiscal 1988 to 1998, consolidated sales increased by 141% at Kyocera, by 94% at Omron, by 71% at Murata, by 186% at Rohm, and by 198% at Horiba.

As shown in Figure 4, their consolidated pretax profits have continued to grow after a decline between 1991 and 1993. Especially, Rohm posted a record consolidated pretax profit in fiscal 1998 for the fifth straight year. Omron boosted its consolidated pretax profit in fiscal 1998 for the fifth straight year. Kyocera hit a new high in fiscal 1997.

The return on equity, a key indicator for investment decisions, has also risen for these companies. The ROE soared by 2.1 percentage points from fiscal 1997 to 16.47% in fiscal 1998 for Rohm, by 0.9 point to 7.80% for Murata and by 0.6 point to 5.55% for Omron. These high ROEs indicate these Kyoto high-tech companies' excellent capital efficiency.

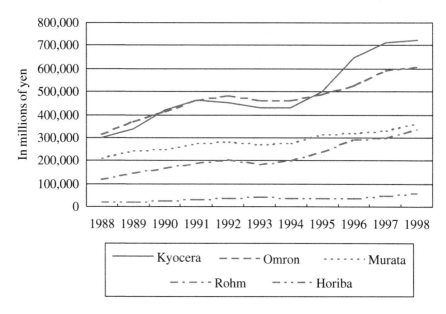

Fig. 3.   Consolidated sales (From *Nikkei Annual Consolidated Corporate Reports 1998*)

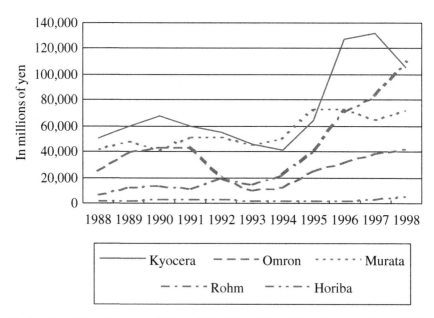

Fig. 4.   Consolidated pretax profit (From *Nikkei Annual Consolidated Corporate Reports 1998*)

Among these companies, Rohm boasted the highest ratio of pretax profit to sales on a consolidated basis at 32.76%, followed by 20.07% for Murata and 14.53% for Kyocera. Their profit ratios are much higher than those for average Japanese companies, representing their excellent profitability.

The Kyoto high-tech companies' excellent earnings performance grows clearer when compared with the earnings of the Japanese electrical machinery industry as a whole in the same period.

After the burst of the economic bubble, the electrical machinery industry saw sales falling by 5% in fiscal 1992 and by 3.2% in fiscal 1993. The sector's pretax profit kept declining substantially for three years. It dropped by 35.7% in fiscal 1991, by 46.9% in fiscal 1992, and by 11.4% in fiscal 1993 (see Figure 5). The Kyoto high-tech companies failed to completely escape from the impact of the bubble burst. Their sales plunged by 7.5% in fiscal 1992. Their pretax profit dropped by about 18% each in fiscal 1991 and 1992, though this was slower than the fall in profit for the sector as a whole. Later, however, the Kyoto high-tech companies boosted sales by more than 10%. Their pretax profit jumped by nearly 50% in two years from fiscal 1994 (see Figure 6).

Overall, both sales and pretax profit for the Kyoto high-tech companies declined more slowly than those for the electrical machinery industry as a whole and recovered faster.

Let us compare the Japanese electrical machinery industry as a whole and the Kyoto high-tech companies in terms of business performance indicators.

Figure 7 illustrates business performance indicators for the electrical machinery industry as a whole and the Kyoto high-tech companies in fiscal 1996. The selected indicators are the sales growth for growth potential, the return on assets for profitability, the capital ratio for safety, the margin of safety ratio for payability, and the value added per person employed for productivity.

The growth potential, or sales growth, for the Kyoto high-tech companies was lower than for the sector as a whole in fiscal 1996, since

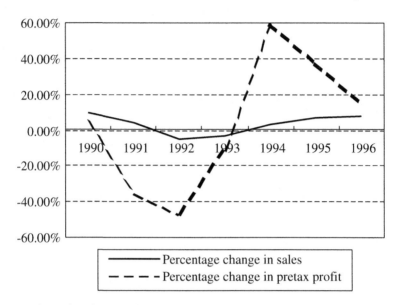

Fig. 5.   Electrical machinery industry (From *Weekly Oriental Economist, Extra Issue: Corporate Financial Carte 1998*)

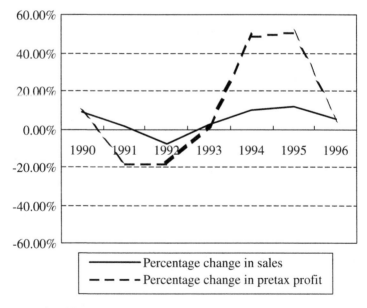

Fig. 6.   Kyoto based high-technology companies (From *Weekly Oriental Economist Extra Issue: Corporate Financial Carte 1998*)

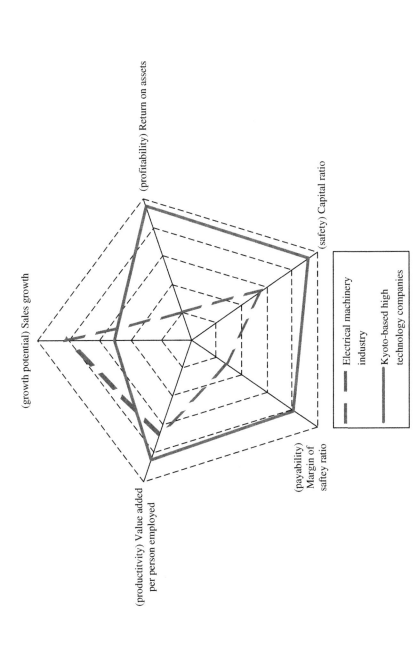

Fig. 7. Comparison between electrical machinery industry and Kyoto-based high-technology companies (From *Weekly Oriental Economist Extra Issue: Corporate Financial Carte 1998*)

the Kyoto firms' sales recovered faster, as noted above. On all the other indicators, the Kyoto firms beat the electrical machinery sector as a whole. Their profitability, or ROA, was nearly five times higher than the sector's, representing their distinctive profitability. Their capital ratio indicating safety was about two times higher and their margin of safety ratio as a payability indicator was about three times higher. The Kyoto firms are thus more stable in business management.

Next, we would like to outline the Kyoto high-tech companies' management systems as the "Kyoto Model."

Top Leadership

Any company depends on the cooperation of labor for its existance. If employees as components of a company have a high morale and produce continuous innovations, the firm will achieve very good earnings performance. A key point in business management is the establishment of a mechanism to enable employees to achieve innovations. Any top business manager must exercise strong leadership in order to take maximum advantage of employees as components of his company, lead to innovations through accumulation and conversion of knowledge, and achieve unpredictable results that are far beyond the initial business goals.

To this end, a top manager who understands core competence must first present his vision specifically and lead based on the vision. If a manager as leader of a company specifies a grand vision in the form of management philosophies, mottos, beliefs, mission statements, or the like and imparts the vision to every employee, the company will have unique organizational characteristics. Such organizational characteristics will be accumulated as knowledge in all aspects of business operations like research and development, production and marketing to trigger innovations and create competitive advantage.

What top managers of Japanese companies do includes (1) materialization of management philosophies, (2) organizational development, (3) preparation of strategies, (4) information activities, and (5) integration of communications (Okumura, *Japan's Top*

Management, p. 154). The reason why top managers of Japanese firms focus their activities on the materialization of corporate philosophies (management philosophies) and organizational development is that any organization into which its top manager has injected values can develop its characteristics and exercise its own strategy-implementing capacity. This prompts the top managers of Japanese firms to maintain and enhance organizational characteristics (Okumura, *Japan's Top Management*, p. 156).

The Kyoto high-tech companies were founded and have developed as technology-developing ventures. Therefore, their founders and their successors as top managers have demonstrated their strong entrepreneurial leadership and endeavored to maintain and enhance their organizations' strategy-implementing capacity based on their corporate philosophies. This allows these companies to make prompt responses to rapid changes in the business environment under their top leadership. Kyocera, Omron, Murata, Rohm, and Horiba under their founders' leadership have differentiated themselves from their rivals by concentrating business resources on their respective core competences and have successfully achieved competitive advantages.

The process can be conceptually illustrated as shown in Figure 8. It is based on a corporate philosophy, which the top manager imparts to employees to heighten the morale of the whole organization. Then, knowledge is actively accumulated and converted in all operations from research and development to production and marketing, allowing innovations to be created organically.

Corporate philosophies here cover corporate values and the like as materialized by mottos, mission statements, business management philosophies, management beliefs, and so on. Employees with high morale can trigger innovations that increase the profitability of a company, allowing the company to achieve higher profit and growth.

Corporate philosophies directly express business goals and values and develop business cultures. The Kyoto high-tech companies give priority to imparting their corporate philosophies to employees. Their top managers are always trying to do so.

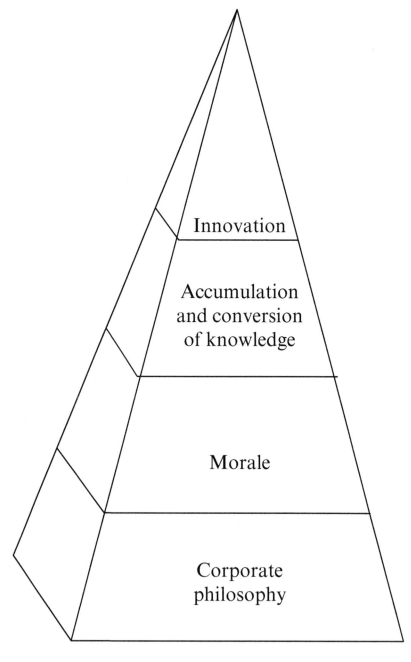

Fig. 8.   Innovation-creating system

The corporate philosophies of these Kyoto firms mostly specify strategic missions as their goals. They reflect their top managers' strong beliefs in business management.

Morale means "working or can-do spirit." As well as each of the employees, the organization as a whole should have a good morale. Any organization with higher morale can enjoy higher performance. Innovation means that good results come about as organizations or their components take maximum advantage of their capacity to improve operations, explore new markets, and develop new products. In innovation, corporate reforms create new value.

The Kyoto high-tech companies possess and improve various systems that boost employees' morale and activate the knowledge accumulation and conversion process to trigger innovation. Supported by corporate philosophy-backed morale, every employee activates knowledge accumulation and innovation, allowing companies to achieve higher earnings and faster growth.

Organization and Control Systems

Generally, in semiconductor and other sectors, whose markets and technologies change fast, earnings are better for companies that adopt a decentralized organization where units have unspecified or vague roles and make horizontal communications frequently (Galbraith and Nethanson, *Strategy Implementation: The Role of Structure and Process*, p. 63). Markets and technologies change drastically in the sectors to which the Kyoto high-tech companies belong. Since their foundation, therefore, they have made use of small units under flat control in organizational management to flexibly respond to drastic market changes under the top leadership. This has allowed them to achieve faster growth.

Such organizational management, or vertical adaptation, may become more difficult as a company grows larger. In this respect, the Kyoto high-tech companies have combined unique managements systems for research and development, production control, cost control, and personnel management to maintain the strength of group dynamics in a bid to develop environments for keeping earnings and growth high.

This management strategy can be illustrated conceptually as shown in Figure 9.

First, these companies utilize strict cost control to increase the efficiency of production, while conducting efficient development of market-oriented products in order to maximize the value added by employees. At the same time, they proactively develop systems and environments that allow employees to innovate.

Accumulation of knowledge and know-how at these companies depends on individuals' or teams' learning based on active internal communications. Therefore, their internal exchange of information is active. They also develop information systems on their own to support active internal communications. Using various internal systems to boost employee morale, the Kyoto high-tech companies increase the likelihood of innovation arising from the accumulation and conversion of knowledge.

Among these internal systems, as shown in Figure 10, are in-house manufacturing of production machines (at Rohm and Murata), process cost control (at Murata), hourly profitability (at Kyocera), self-supporting units (at Murata), in-house awards (at Rohm and Horiba), "meisters" or experts (at Horiba), annual salaries (at Horiba and Rohm) and new business development centers. The Kyoto high-tech companies utilize such systems to boost their employees' morale.

The internal systems also include ingenious ones like the amoeba-based management system (at Kyocera), matrix management (at Murata), and the product-oriented management system (at Horiba). These ingenious systems are the key to boosting employees' morale for knowledge conversion and innovation.

In the pursuit of business management efficiency, the Kyoto high-tech companies have adopted smaller organizations to boost employees' morale and smaller units to control for stricter cost savings.

In the present uncertain market environment, the decentralization of an organization can boost management efficiency. This

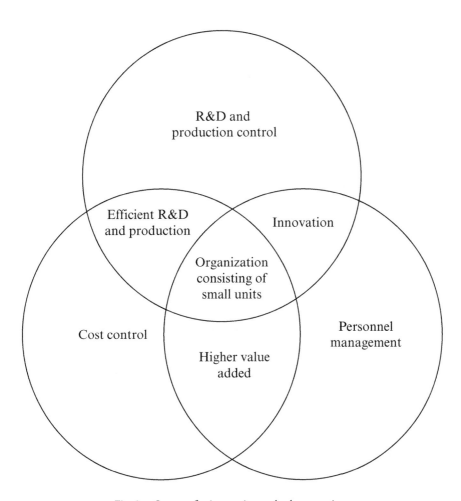

Fig. 9.   Systems for innovation and robust earnings

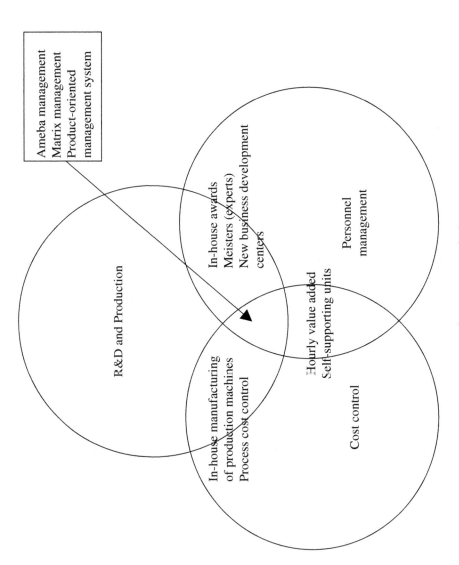

Ameba management
Matrix management
Product-oriented
management system

In-house awards
Meisters (experts)
New business development
centers

Personnel
management

R&D and Production

In-house manufacturing
of production machines
Process cost control

Hourly value added
Self-supporting units

Cost control

Fig. 10.   Systems for innovation and robust earnings

is especially true in the electronics industry, which features fast technological innovation. Technology development-oriented companies must repeat in-house communications frequently in order to understand market needs promptly and to link them to product development. In this sense, the Kyoto high-tech companies have decentralized their respective organizations to increase employees' consciousness of participation in business management and activate communications. They have devised ways to induce employee innovation.

In production control, these companies have developed unrivaled production technologies to achieve a competitive advantage. Such production technologies include integrated production of raw materials and finished products (at Murata), and in-house manufacturing and maintenance of production machines (at Rohm and Murata).

Philanthropy

Kyocera, Omron, Murata, Rohm, and Horiba have grown with support from various people in Kyoto, a society featuring a strong sense of comradeship. Appreciating such support from local people and communities and acknowledging their social significance, they have been active in philanthropy. Kyocera's Kazuo Inamori has founded the "Kyoto Award" to honor personalities for their life work. Omron has positively participated in social welfare programs. Murata has supported academic research. Rohm has created the Rohm Music Foundation. Masao Horiba has contributed to developing venture businesses.

Japanese shareholders, unlike their U.S. counterparts, had generally refrained from putting pressure on companies to thoroughly pursue capital efficiency. This had left Japanese business managers free from considering accountability to shareholders.

But the Kyoto high-tech companies have traditionally given priority to accountability. They have always given full consideration to investor relations. This is a business management policy that they have learned through their overseas experiences, including competition in overseas markets and listings on overseas stock markets. Apart from robust earnings, such an accountability policy is the reason why these companies enjoy a good reputation in the stock market at present.

The Kyoto Model is a business management model designed to harmonize the sustained growth of companies with society.

In conclusion, the Kyoto high-tech companies' management systems feature the following, including philanthropy:

- Strong top leadership
- Small organizational units to control
- Innovation-inducing personnel management
- Strong cost consciousness
- Market-oriented research and development
- Advanced production technologies
- Making use of stock markets and emphasizing investor relations
- Emphasizing philanthropy

Chapter 2

Why Were the High-Tech Companies Founded in Kyoto? — Naturally or Accidentally?

We would like to consider why the Kyoto high-tech companies, as R&D-oriented venture businesses, were founded in Kyoto and developed into global firms.

Analysis Based on Macro Data

As noted in the Introduction, Japanese-style business management has allowed various kinds of knowledge to be accumulated within an organization. This has been a key feature of Japanese-style management and the base for Japanese manufacturing companies' competitive edge. Especially, knowledge and know-how accumulated at the point of production have contributed much to allowing Japanese manufacturers, including processing and assembly companies like automakers, electrical machinery builders, and electronics manufacturers, to establish their competitive edge.

In steel, TV sets, VCRs, cars, semiconductors, and other product categories in which Japanese firms have been relatively competitive, they have made various innovations in production machinery, technologies, and systems, which they have linked to their competitive edge (Kagono *et al.*, *Comparison between Japanese and U.S. Business Management Styles*, p. 97). Especially, the way Japanese companies accumulate business resources (knowledge) has been useful for organized learning in microelectronics, the center of technological innovations where technologies have developed in line with learning

27

through experiences, applications, and diffusion (Imai and Komiya, *Japanese Companies*, p. 11).

Let us analyze the electrical machinery market based on macro data. As indicated in Figure 11, the Japanese electrical machinery industry has persistently achieved substantial growth in shipments and exports of products in value. The growth has been far faster than that for cars and other transportation machines, and steel. Electrical machinery has continued to exceed transportation machinery in shipment value since 1983, indicating that the "electronization" since the 1980s has supported the growth of the electrical machinery industry.

In value added, the electrical machinery industry has outperformed the transportation machinery and steel sectors since the beginning of the 1980s. This industry's value added has maintained substantial growth (see Figure 12).

The electronics components industry, including the Kyoto high-tech companies, has served as the base for supporting electronization. This industry has grown in line with progress in electronization. As indicated in Figure 13, electronics components exports, including that of semiconductor computer chips, have increased substantially in line with electrical machinery exports.

A typical example of this relationship has been the Japanese manufacturers' domination of the global VCR market. Only the Japanese electronics components industry has been able to supply a huge quantity of quality components for VCRs at low cost.

In the 1970s, the electrical machinery industry became Japan's industrial leader. Behind this development, the electronics components sector was the foundation (Itami, *Three Waves for Japanese Industry*, p.158). Supporting the Japanese electrical machinery manufacturers' domination of the global VCR market was Japan's electronics components sector. Even at the initial R&D stage, components manufacturers proactively make cost-cutting and quality-improving proposals to and cooperate with machinery producers. As such, cooperative relationships continue over a long period of time, the know-how for the production of unit components shifts

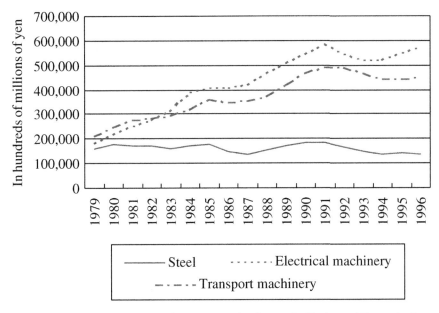

Fig. 11.   Manufactured products shipments in value (From *Weekly Oriental Economist Extra Issue: Economic Statistics Almanac 1998*)

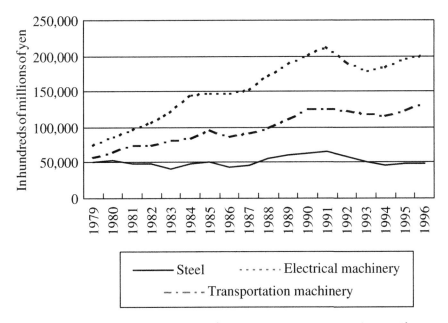

Fig. 12.   Valued added (From *Weekly Oriental Economist Extra Issue: Economic Statistics Almanac 1998*)

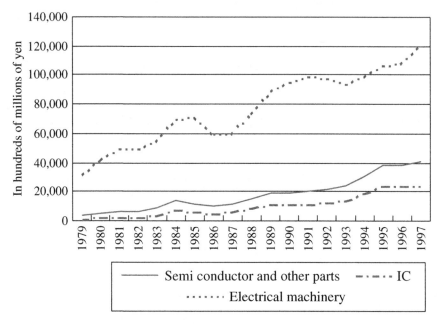

Fig. 13.   Customs-cleared exports in value (From *Weekly Oriental Economist Extra Issue: Economic Statistics Almanac 1998*)

from the machinery manufacturers to the component manufacturers (Nakazawa, *New Era for Small Companies*, p. 171).

Japanese component manufacturers have apparently taken advantage of technology accumulation based on the Japanese style of business management and the accumulation of know-how through joint product development with electrical machinery and electronics manufacturers to produce synergy effects and establish the foundation of Japan's electrical machinery industry. In this sense, without Japanese components manufacturers, Japan's electrical machinery producers could have failed to dominate the global market. The component manufacturers' own growth has depended heavily on the electrical machinery producers. Component and machinery manufacturers might have cooperated in accumulating technologies.

In the analytical instrument area where Horiba has been active, production based on orders given by specific customers has been dominant. Therefore, any company in the area is required to develop excellent technologies for all the production, marketing, and maintenance stages. Accumulation of knowledge thus plays a key role in allowing companies to acquire a competitive edge.

As indicated in Figures 11 and 12, the market for electrical machinery manufacturers has expanded substantially on progress in motorization and growing interest in environmental problems. As a result, electronics component manufacturers have successfully accumulated technologies and knowledge through joint production, marketing, and maintenance with customers.

Kyotoism

The market environment has thus supported the growth of the Kyoto high-tech companies. But the question is: why have these companies originated from Kyoto?

While these high-tech companies were founded in Kyoto, their founders other than Masao Horiba were not from Kyoto. This may indicate that the concentration of R&D-oriented ventures in Kyoto has only been accidental. Given the historical background and the contemporary backdrop of their founding, however, the concentration cannot necessarily be interpreted as accidental.

Before the Meiji Period, Kyoto was Japan's capital and home of the Emperor. Culture and traditions flourished. Various industries had prospered and raced to develop skills and technologies. Behind the development of various industries in the Japanese capital might have been Japanese people's traditional respect for craftsmen.

In this respect, writer Ryotaro Shiba says:

Such reputation was the purpose in life for craftsmen and supported by the Japanese people who favor craftsmen. Many Japanese craftsmen, including the central characters in comic stories, pursued such unrewarded honor as their purpose in life. Frank Gibney, an expert on Japan, has described Japan as "the country that respects craftsmen." This description represents an original feature of Japanese society. (Shape of This Country 2, p.180).

The cultural respect for craftsmen has been deeply rooted especially in Kyoto. This background has apparently played a great role in allowing high-tech companies to be founded in Kyoto. Kyoto high-tech companies are mostly component manufacturers. The multilayer traditional industries and craftsmanship developed in Kyoto might have contributed much to the development of microtechnologies for components development and production and to making up for shortages in production technologies.

In Kyoto featuring such traditional craftsmanship, there may be many martinets who are too proud. Some may be concerned that this could make it difficult to launch businesses in Kyoto. In fact, Kyotoites are generally viewed as too proud. But such pride might have served to eliminate second-rate technologies or products, leading to a favorable outcome such as the development of high-quality products by the Kyoto high-tech companies.

In Kyoto filled with traditional businesses, any commercial activities that affect their turfs may be faced with resistance and interference. Therefore, anyone who launches a business in Kyoto must do something unique. This background might have led to very creative companies that do not imitate others. The climate that fosters creativity may be behind the fact that most of the Nobel Prize winners in Japan are from Kyoto University.

Kyotoites and the Kyoto community, which puts much emphasis on traditions, have assessed everything very strictly. Such strict assessment might have served to improve the technologies and products of companies founded in Kyoto, allowing them to develop into world-class ones.

Sustainable Business Management

Kyotoism that is rather conservative and steady has had a great impact on the Kyoto high-tech companies' management systems. Kyotoism has led these companies to commonly feature steady management that is based on corporate philosophies and focused on core competences. The steady management is reflected in their strict cost control systems. This has led these companies to keep to their optimum size with priority given to customer satisfaction, instead of expanding their size for mass production or mass marketing.

Instead of expanding shares of mass markets, the Kyoto high-tech companies have strategically chosen to win top market shares of niche markets and achieve robust earnings by accumulating new functions and cost-cutting efforts for the longer survival of their products. Rather than trying to beat rivals and win in competition, they have tried to improve their products and production technologies only for their own survival or permanent presence. This has allowed them to acquire and build up their competitive advantage. As a result of their continual self-improvement, they have successfully beaten rivals and acquired top market shares. This may be termed sustainable business management.

Such a business management style may look very fresh to Japanese companies after the bubble economy period. But this style is frequently seen in traditional family businesses. In this sense, this is a management style that current Japanese companies have forgotten. The Kyoto high-tech companies might have naturally developed their management style in their growth in the Kyoto climate.

Another reason R&D ventures have been born in Kyoto is that there are a large number of universities and other educational institutions in Kyoto.

As many universities have been established in Kyoto since the Meiji Period, academic facilities have been abundant and academic levels have been kept high. The presence of a large number of universities in Kyoto has been closely linked to the Meiji Restoration. When Kyoto lost its vitality with the relocation of the Japanese capital from Kyoto to Tokyo, Kyoto Prefecture Governor Masanao Makimura launched industrial, educational and other development promotion measures. Kyoto's modernization began with educational system development including the introduction of a school district system. The abundance of universities in Kyoto dates from the education system development in the early Meiji Period, including the establishment of universities.

In Kyoto, highly educated people were available for employment, and industry–academia cooperation in research made progress. Such a climate might have been the driving force behind the emergence of R&D-oriented high-tech ventures like Omron, Murata, and Horiba. But the industry–academia cooperation has declined since the rise of the student movement.

But even the presence of leading academic institutions failed to prevent highly educated people from moving from Kyoto to business centers like Tokyo, Osaka, and Nagoya. As discussed in Chapters 5 and 6, however, the Kyoto high-tech companies have overcome this problem by devising various measures to stimulate employee morale. They have taken maximum advantage of the limited business resources in Kyoto to develop into world-class manufacturers.

Absence of Big Banks and Companies

The absence of big banks has resulted in a weak banking sector in Kyoto. This might have allowed industrial companies in Kyoto to gain more freedom in business management.

But they have had difficulties raising funds from banks. R&D-oriented high-tech ventures like Kyocera, Omron, Murata, Rohm, and Horiba have been difficult for banks to assess for lending decisions. The difficulties might have prompted these ventures to look to, and exploit, the capital market. This might have led to their successful penetration into overseas markets, allowing Kyocera, Omron, and Murata to offer their equity shares on overseas stock markets.

As well as big banks, big companies like the *zaibatsu* conglomerates in Tokyo and Osaka have been absent in Kyoto. This fact might also have had some impact on the management of the Kyoto high-tech companies. In the absence of control by big companies as well as banks, these ventures, even though starting as subcontractors, might have been able to freely develop into leading companies.

The Kyoto climate has brought about the spirit of independence and self-respect, prompting top managers of these ventures to resolve that they should never end up as subcontractors. Through transactions with big companies in Tokyo, Osaka, and the United States, these ventures might have accumulated know-how and built up their competitive edges gradually.

Chapter 3

Real Faces of Kyoto High-Tech Companies: History of Growth Since Foundation

Core-Based Growth

The five firms cited as Kyoto high-tech companies are all R&D-oriented ventures founded by engineers. Each features original technologies and products at the core. The cores are fine ceramics for Kyocera, control systems for Omron, electronic components for Murata, custom ICs for Rohm and analytical instruments for Horiba. They have built on their respective core products to develop other products for their expansion.

Let us review the history of how the Kyoto high-tech companies developed their core products for their founding and have continued their growth as shown in Table 4.

Kyocera: From Ceramic Components Manufacturer to Integrated Circuits Manufacturer

Kyocera began as Kyoto Ceramic Co., founded by Kazuo Inamori and some colleagues in 1959 after they quit another company. Kyoto Ceramic was then a fine ceramics manufacturer capitalized at 3 million yen with 28 employees. Before founding Kyoto Ceramic, Inamori had been in charge of research and development, production and marketing for new ceramics at another company. He had already shone as an able, charismatic business manager by developing his division at the company into its growth center. Inamori quit the

Table 4.   Outline of Kyoto high-tech companies

	Kyocera	Omron	Murata	Rohm	Horiba
Year of founding	1959	1933	1950	1954	1953
Founder	Kazuo Inamori	Kazuma Tateisi	Akira Murata	Kenichiro Sato	Masao Horiba
Products	Fine ceramic components, semiconductor components, electronic components, communications equipment, information equipment	Control systems, electronic payment machines, public information systems, healthcare equipment	Capacitors, resistors, voltage products, coil products, circuit designs	Integrated circuits, semiconductor chips, passive components	Chemical measurement systems, electronic information equipment, engine measurement instruments
Market share	62% for fine ceramics components, 46% for semiconductor packages, 42.9% for solar cells, 10.6% for mobile phones (in Japan)	40% for control systems, 12.2% for communications relays (in Japan)	50% for chip monolithic capacitors, 80% for ceramic filters, 80% for ceramic resonators (in the world)	42.2% for small signal transistors, 26.2% for chip resistor, 28.0% for semiconductor lasers (in Japan)	80% for engine measurement instruments, 40% for pH meters, 30% for flue gas analyzers (in Japan)
Workforce (1998)	13,594	7,154	4,489	2,542	1,074
Unconsolidated sales (1998)	491,739	432,713	290,420	272,839	29,443
Consolidated sales (1998)	725,312	611,795	362,252	335,922	62,425
Unconsolidated pretax profit (1998)	65,737	21,076	30,488	61,352	2,225
Consolidated pretax profit (1998)	105,380	42,243	72,694	110,064	5,464
Ratio of consolidated pretax profit to sales (1998) (%)	14.53	6.90	20.07	32.76	8.75
Unconsolidated ROE	5.57	4.01	5.86	10.96	2.94
Consolidated ROE	6.36	5.55	7.80	16.47	4.27

Table 4. (*Continued*)

	Kyocera	Omron	Murata	Rohm	Horiba
Year of initial public offering	1971	1962	1963	1983	1971
Exchanges for listing	Tokyo, Osaka, Kyoto, New York	Tokyo, Osaka, Kyoto, Nagoya, San Francisco	Tokyo, Osaka, Kyoto, Singapore, San Francisco	Tokyo, Osaka, Kyoto	Tokyo, Osaka, Kyoto

Notes: Sales and pretax profit are in millions of yen. Consolidated earnings are based on the U.S. Securities and Exchange Commission guidelines for Kyocera, Omron, and Murata.
Sources: Data from the five companies; *Market Shares '98*; *Nikkei Business Data*, January 1998; *Nikkei Business*, May 11, 1998; *The 21*, May 1998; *Japan Market Share Handbook 1998*; *Japan Company Handbook*; etc.

company due to an internal dispute and invited some colleagues to help found Kyoto Ceramic, which has developed into Kyocera. He was 27 years old.

Kyocera initially manufactured ceramic insulators for television sets mainly for delivery to electric appliance manufacturers. But leading Japanese electronics manufacturers were wary of dealing with Kyocera, which was then a small, little-known company. This prompted Inamori to go to the United States in pursuit of a market only a few years after founding Kyoto Ceramic.

Kyocera was faced with a lot of problems and difficulties in exploring the U.S. market. But its strenuous and continuous efforts bore fruit. In 1965, it successfully landed orders for resistor rods for the Apollo Program from Texas Instruments. This success led Kyocera to deal with not only other U.S. companies but also Japanese firms that had heard of its good reputation.

In the 1960s, when the semiconductor industry suddenly developed, Kyocera began to receive requests from Silicon Valley companies for the production of ceramic semiconductor components. In 1968, it accepted a request for the development of multilayered ceramic packages for integrated circuits (ICs) which have become its core product. Kyocera was then asked to develop the package in only three months.

Soon after accepting the request, Kyocera found that it had no machine to manufacture not only the product but also its ceramic material. Even under such extreme difficulties, Kyocera engineers made superhuman efforts. As a result, they successfully solved the various technological problems one by one and fabricated the product. The development of multilayer IC packages was later responsible for Kyocera's fast growth.

Kyocera now manufactures products ranging from semiconductor packages for the guts of computers, various electronic components indispensable for communications and other equipment, and fine ceramic components for industrial machines and automotive engines to popular products including mobile phones, printers, cameras, and solar home power generators. Behind Kyocera's wide expansion of its product range was not only its own product development efforts but also its aggressive mergers and acquisitions strategy.

Kyocera merged with Cybernet Electronics Corp. in 1982 and with Yashica Co. in 1983, developing from a components maker into a diversified manufacturer covering information, communications, optical, and other final products in addition to components. In 1989, Kyocera acquired Elco Corp. to participate in the market for connectors for electronics. In 1990, it absorbed AVX Corp. to become a global capacitor manufacturer. In this way, Kyocera has steadily solidified its position as a comprehensive electronic components manufacturer.

Omron: Grasping Social Needs

Omron was founded as Tateisi Electric Manufacturing Co. by Kazuma Tateisi in 1933. Before founding Tateisi Electric Manufacturing, he had worked for an electrical machinery manufacturer and had managed Saikosha, which manufactured and marketed trouser presses and knife grinders.

Saikosha was not expected to change for the better. But a turnaround came as Tateisi considered commercializing a faster X-ray timer and successfully developed the timer. X-ray timers were then mechanical and unsuitable for taking grab shots including an X-ray picture of the chest. The medical world then demanded more accurate timers. Upon the successful development of the faster X-ray timers, Saikosha was renamed Tateisi Electric Manufacturing.

Tateisi Electric Manufacturing was initially producing the X-ray timer on an original equipment manufacturing basis and delivering various connectors to switchboard and heavy electrical machinery manufacturers. Just before the Pacific War, Tateisi Electric Manufacturing received a request from the Aeronautics Laboratory of the University of Tokyo to develop the precision switch that had already been used for aircraft in the United States. Even just before the war, the company obtained catalogs and samples from the United States and built on them to accumulate experiments. Learning by trial and error, the firm successfully developed the precision switch. Although the precision switch failed to be commercialized during the war, this technological breakthrough led to the development of automation systems and had a great impact on the fate of the company.

In and after 1955, electric home appliance manufacturers began to use automation systems. As the whole of the industrial world grew interested in automatic control systems, Omron took the initiative in developing and merchandising precision switches, magnet relays, time relays, and other automation products.

Under Kazuma Tateisi's "Social Needs Theory" calling for "anticipating social needs and filling them with pioneering technology," Omron later took advantage of its automatic control technologies for developing and manufacturing ticket vending machines, traffic control systems, automatic ticket checkers, and automatic cash dispensers. Especially, automatic cash dispensers were later connected on-line to central computers to create the world's first on-line cash dispenser network.

Murata: Focusing on Ceramics

Murata Manufacturing Co. originated from a small backstreet factory that was run by the father of Akira Murata. The factory was manufacturing insulators. After taking over the factory from his father in 1944, Akira Murata received a request from a large electrical machinery manufacturer for the development of titanium capacitors (early-stage ceramic capacitors). Unlike the insulators that his factory had produced, the product was an electronic component. After three months of learning by trial and error based on written and oral information from experts, Murata successfully developed the titanium capacitor.

Ceramic capacitors including the titanium capacitor have become Murata's mainstay. As a small company then, Murata survived the turbulent war period by delivering the titanium capacitor to the large electrical machinery manufacturer.

Even amid the confusion just after the war, Akira Murata continued aggressive R&D operations, including development of epoch-making materials like barium titanate, building the foundations of Murata Manufacturing's later growth. The company accumulated R&D operations and improvements for ceramic capacitors and explored new markets for ceramic capacitors for television and stereo sets. In 1960, Murata became the first Japanese manufacturer to sell products to big U.S. companies like Motorola Inc. and General Electric Co.

In 1962 after three years of R&D operations, Murata successfully developed small ceramic filters. Sony Corp. adopted the filter for transistor radio receivers. In 1971, General Motors Corp. introduced the filter for car radio receivers. Murata then expanded both the Japanese and the overseas markets for the small ceramic filters. Murata now commands 80% of the global market for small ceramic filters.

Murata's electronics components, though without attracting widespread attention, are widely used for popular electrical appliances ranging from color TVs, video cameras, mobile phones, and laptop personal computers to refrigerators and air conditioners.

Rohm: Bold Challenging Approach on ICs

Rohm's predecessor was Toyo Electronics Industry Corp., which started in 1954 to produce a small resistor that Kenichiro Sato had devised and patented when he was a university student. As an electric appliance boom then developed, the small resistors produced by

Toyo Electronics Industry sold very well for portable transistor radio receivers. As a result, the firm's earnings expanded fast.

Toyo Electronics Industry then reached a turning point. Kenichiro Sato, who had made big profits on small resistors, courageously shifted to integrated circuits as his firm's next-generation mainstay. At that time ICs were still relatively unknown, just after their development in the United States in 1967. Prompting Sato to make the shift was a sense of crisis that ICs would replace resistors, as well as his entrepreneurial spirit that encouraged him to explore an unknown market.

In the face of difficulties in finding engineers who were well versed in ICs, Sato suddenly decided to expand into Silicon Valley. As early as 1971, Rohm founded Exar Corp. as an IC development base in Silicon Valley. This was the first firm that a Japanese company established in Silicon Valley. Exar made its initial public offering on the Nasdaq market in 1985.

Rohm had borrowed an enormous amount of money from banks for its expansion into the United States, but it had difficulties in developing ICs and its earnings deteriorated during the subsequent oil crisis. Rohm then came to the brink of bankruptcy. Its investment finally began to bear fruit around 1977 when sales of its custom ICs exploded, allowing the company to repay 10 billion yen in borrowings in only three years.

Later, Rohm continued to grow while producing mainly niche products including custom ICs and special memories that large companies hesitated to manufacture. Rohm kept away from the development or production of dynamic random access memories and other cutting-edge computer chips, where sophisticated technologies and massive investment are required. While leaving such products to be produced by large companies, Rohm manufactured older-generation ICs adapted to customers' specific orders. Avoiding direct competition with large companies, it has successfully captured customers.

Rohm now boasts top shares of the resistor and semiconductor laser markets in Japan. It also commands a dominant share of the global market for motor control ICs, used for floppy disk drives.

Horiba: Originating from a Student Venture

Masao Horiba founded the Horiba Radio Laboratory as the predecessor of Horiba Ltd. in 1945, when he was a junior at university. The laboratory was initially engaged in research and development of electric circuits. In fact, it managed to survive by offering repair services for electric home appliances.

Since electric circuit problems were then mostly traced to capacitors, Masao Horiba made up his mind to develop high-quality capacitors. Depending on loans and learning by trial and error, he finally completed prototypes, but the inflation caused by the Korean War forced him to give up the commercialization of high-quality capacitors in which massive investment was required. Instead, Horiba chose to manufacture and market the pH meters that he had developed for making capacitors.

In 1953, he founded Horiba Ltd. with personal investment he requested from J. Osawa & Co. chairman Yoshio Osawa and Keifuku Electric Railroad Co. president Yoshijiro Ishikawa. This might have been one of the first cases of venture capital investment in Japan. In appreciation of the key investors, Horiba has established in-house prizes bearing their names. The Osawa Prize is awarded to employees who have made distinguished achievements in marketing. The Ishikawa Prize is designed for those with outstanding management achievements. In addition, Horiba has had several other prizes bearing the names of its mentors.

Horiba Ltd. successfully developed the first ever pH meter in Japan, building the foundation of its measurement instrument business. A turning point came as employees developed an automobile emission analyzer without any permission from Masao Horiba. Later, the automobile emission analyzer became a big hit, accounting for as much as 40% of Horiba's sales and an 80% share of the global market.

Horiba took the lead in research and development of automobile emission analyzers in the world as various emission regulations were introduced in response to atmospheric pollution problems in the

1970s and environmental problems in the 1980s. The Horiba group as a whole now commands 80% of the global market for measurement instruments.

Horiba has continued to focus its business resources on analytical technologies, developing various analyzers, including numerous analyzers for semiconductor chips, scientific measurement instruments for research into new materials and energy sources, truck driving control systems, medical diagnostic instruments, and atmospheric and water pollution monitoring systems. High-technology development and growing interests in environmental conservation have allowed Horiba to expand its market share more and more.

Keeping Top Market Shares for Core Products

The Kyoto high-tech companies have given priority to their respective core technologies and focused their business resources on improving core technologies and developing unique products. Their technology and product management cannot be imitated by others. As a result, they command dominant market shares worldwide as well as in Japan.

In the sectors to which the Kyoto high-tech companies belong, any company that overcomes the severe competition and acquires the top market share can take advantage of the dominant share to achieve high earnings while others have difficulties in penetrating the market. Generally, the products of the Kyoto high-tech companies must be diversified for limited production runs and feature high quality and low prices. Investment in production equipment for these products is costly while these products are low priced. Companies manufacturing these products cannot make a profit unless they command a high market share. The markets for the Kyoto high-tech companies are, therefore, protected by high barriers to entry. It is difficult for others to penetrate into these companies' markets.

The Kyoto high-tech companies' market shares are as follows:

Kyocera whose main products are electronic components has captured 62% of the fine ceramics component market and 46% of the semiconductor package market. These percentages represent top market shares. The firm also commands 42.9% of the solar cell market, keeping a long distance ahead of rivals. Kyocera has acquired the third largest share, 10.6%, of the mobile phone market by producing phones mainly for the DDI Cellular Group that was founded primarily by Kyocera.

Omron has obtained 40% of the Japanese market for control systems that are the firm's mainstay.

Murata commands about 50% of the global chip monolithic capacitor market. Chip monolithic capacitors are its core product accounting for one-third of its total sales. They are used for various electronic appliances and information systems, including color TVs, VCRs, and video cameras. Murata has also obtained as much as 80% of the global market for ceramic filters, which are indispensable for mobile phones whose market has been expanding rapidly. It accounts for 80% of the global market for ceramic resonators, which are used for color TVs, VCRs, and video cameras. Murata thus boasts leading global market shares for a range of products.

Rohm has top global market shares for custom ICs, small signal transistors, chip resistors, and semiconductor lasers. Custom products account for nearly 70% of these products manufactured by Rohm. Custom ICs, used mainly for personal computers' peripheral systems, have a fast-growing market. Custom ICs are niche products that are difficult for large companies to handle, since flexible engineering capacity to meet customers' specific orders is required for their production. They thus secure high profit margins and stable revenues.

Horiba commands some 80% of the global market for automobile emission analyzers and top shares of regional markets in Europe as well as Japan and the United States. In some countries, its share exceeds 90%. The automobile emission analyzers measure nitrogen oxide, carbon oxide, and other pollutants in exhaust gas and are used in automakers' development and testing processes. As for the pH meters that Horiba has produced since its foundation, the company still accounts for 40% of the Japanese market. Aided by global moves to toughen automobile emission regulations and by growing consciousness about environmental conservation, the market for analyzers, including atmospheric and water pollution monitoring systems, has been expanding ever wider.

Aggressive Globalization

The Kyoto high-tech companies commonly became famous first in the United States and other foreign countries, rather than in Japan. They expanded overseas almost simultaneously. Their deals with leading U.S. companies have allowed these once unknown firms to boost their creditworthiness in the Japanese market. They advanced into the United States in or after the second half of the 1960s, laying the foundation for overseas expansion. In the United States, Rohm founded a subsidiary, Horiba formed strategic partnerships, Murata sold products to General Motors, and Kyocera won orders from Texas Instruments.

Kyocera expanded transactions with leading U.S. companies on the strength of orders from Texas Instruments for resistor rods designed for the Apollo Program. This allowed Kyocera to increase deals with large companies in Japan.

In the 1970s, Kazuo Inamori was confident of the promising future of large-capacity composite capacitors using the multilayered ceramics technology. Kyocera then signed an agreement with Aerovox Corp. to produce multilayered ceramics capacitors in Japan for global sale under the U.S. firm's license. Aerovox was the predecessor of AVX Corp., which Kyocera later acquired.

In 1990, Kyocera acquired AVX through an equity swap deal, becoming a global capacitor manufacturer. Kyocera was the first Japanese firm to use the equity swap. AVX, as a Kyocera subsidiary, was listed on the New York Stock Exchange in 1995.

Omron developed an automatic cash dispenser at the request of a U.S. company. The machine was delivered in 1965. For the Japanese market, Omron developed off-line cash dispensers before launching the world's first on-line cash dispensers. In the 1970s, Omron entered the United States for the production and marketing of desktop electronic calculators, but it later withdrew from the business.

In 1989, Omron built a four-pole (Japan, the United States, Europe, and Asia) system to control global operations, establishing regional head offices in the United States, the Netherlands, and Singapore.

Murata started marketing in the United States in the 1960s. As Murata was then thriving sufficiently in Japan, some within the company raised objections to the risky expansion into the United States. But Akira Murata anticipated the promising future of the large U.S. market with high technological levels and traveled to the United States almost every year to market the mainstay ceramic capacitors.

As a result, Murata became the first Japanese company to successfully sell ceramic capacitors to leading U.S. firms such as Motorola and General Electric. By around 1965, exports had increased to one-fourth of Murata's total ceramic capacitor sales. Murata founded a U.S. subsidiary in 1965.

In a bid to launch its new IC business, Rohm directly advanced into Silicon Valley. In 1971, it became the first Japanese company to have a subsidiary in Silicon Valley. After launching IC production in Silicon Valley, Rohm established subsidiaries in West Germany, South Korea, Brazil, Hong Kong, and Singapore from the late 1970s to the 1980s to enhance global production and marketing operations.

Horiba activated its overseas business strategy in 1968. It founded a joint venture with a U.S. venture business in 1970 for the marketing of exhaust gas analyzers. The joint venture took advantage of California's toughening exhaust gas regulations to grow steadily. In 1973, Horiba terminated the joint venture agreement and founded a wholly owned subsidiary headed by an American president to enhance its U.S. business strategy.

Horiba later opened offices in Switzerland, Britain, France, Germany, Singapore, and other foreign countries to put its overseas business strategy into orbit. Especially in Europe, Horiba has 80% of the regional automobile emission analyzer market. In some European countries, its market share exceeds 90%. Europe has become a key market for Horiba and the company has delivered these analyzers to almost all European automakers including Mercedes-Benz, Volkswagen, Renault, and Fiat. This may be because Horiba opened European offices as early as 1970 to establish its European marketing, services, and R&D bases.

This history indicates that the Kyoto high-tech companies took advantage of their early successful expansion into the United States and other overseas markets to grow in Japan. Their overseas expansion triggered their growth.

The large size and depth of the U.S. market are cited as the reasons why R&D-oriented ventures like the Kyoto high-tech companies have successfully expanded into the United States. The U.S. business climate has allowed large companies to fairly assess and adopt products of even unknown small firms based on technology. In Japan, large companies have refrained from adopting the products of unknown firms even if the products are excellent.

The Kyoto high-tech companies have had financing strategies that conventional Japanese companies do not have. Their difficulties in raising funds in Japan have prompted them to offer their equity shares in overseas markets earlier than other Japanese companies. They have raised funds from overseas stock markets.

Murata became the first Japanese firm to be listed on the Singapore stock market in 1976. It later made a debut on the San Francisco Stock Exchange. Kyocera has been listed on the New York Stock Exchange and Omron on the San Francisco exchange. The Kyoto high-tech companies have thus featured global fund-raising operations.

As a result, these companies have many foreign shareholders. Foreign ownership is as high as 26.0% for Kyocera, 29.5% for Omron, and 36.2% for Murata. The largest shareholder in Murata is a foreign institutional investor. Rohm has not been listed on any overseas stock market, but its excellent profitability has attracted foreign investors. Its foreign ownership is even higher at 41.7%. The high foreign ownership percentages have prompted these companies to positively disclose corporate information to investors.

Expansion through M&A Deals

The Kyoto high-tech companies have positively exploited M&A deals to enhance their core competences or launch new businesses, as indicated in Table 5.

Their M&A deals have consistently been based on a constrained diversification strategy to maintain and enhance competitiveness in markets for their core product categories, rather than on a diversification strategy to move into unfamiliar markets.

Aggressive, Bold Acquisitions — Kyocera

Kyocera began as a manufacturer of fine ceramics components, but it has developed into a diversified manufacturer producing products ranging from semiconductor packages, electronic components for communications equipment, and various fine ceramics components for industrial machines and automobile engines to finished products like mobile phones, printers, cameras, and solar home power generation systems. A major factor behind the development has been a series of aggressive M&A deals both at home and abroad.

Table 5.   Kyoto high-tech companies' M&A strategies

Company	Year of acquisition	Acquisition target	Nationality	Mainstay products
Kyocera	1982	Cybernet Electronics	Japan	Information and communications equipment
	1983	Yashica	Japan	Optical equipment
	1989	Elco	U.S.	Connectors
	1990	AVX	U.S.	Capacitors
Omron	1991	Nippon Data General	Japan	Minicomputers
Murata	1980	Erie Technological Products	U.S.	Electronic components
Rohm	1986	Excel	U.S.	ICs
Horiba	1996	ABX	France	Blood cell counters

In Japan, Kyocera acquired an equity stake in Cybernet Electronics Corp. in 1979 and merged with it in 1982 to add final information and communications equipment products to its product lineup. In 1983, Kyocera merged with Yashica Co. to add fine optical products to the lineup. Through these mergers, Kyocera shifted from being an electronic component manufacturer to becoming a diversified manufacturer of components and finished products.

Cybernet's information and communications equipment technologies have been especially indispensable for the later development of mobile phones and other information equipment. Yashica's optical technologies have been necessary for developing copiers, compact disk players, and other products. Given the present rapid growth of the information and communications equipment market, we can find that these mergers have been of great significance to Kyocera.

Kyocera has also carried out bold acquisitions overseas to solidify its position as a comprehensive electronic component manufacturer and expand its overseas development, production, and marketing bases. In 1989, Kyocera acquired Elco Corp., a major electronics connector manufacturer. With five production bases in Japan, Europe, and the United States, Elco was delivering connectors to Sony Corp. and other leading companies. The Elco acquisition allowed Kyocera to make up for a deficiency and obtain new marketing channels in Japan, Europe, and the United States.

In addition, Kyocera acquired AVX Corp., a global capacitor manufacturer, in 1990. Kyocera thus obtained the capacitor division that it had lacked, as well as 17 production bases and marketing channels in the world. AVX has been indispensable for Kyocera's globalization as a comprehensive electronic component manufacturer.

The AVX acquisition was different from the Elco buyout in that Kyocera exploited the equity swap method. Kyocera was the first Japanese company to use an equity swap for an acquisition. While acquisitions in Japan are usually made through cash payments, Kyocera was able to acquire AVX through an equity swap deal, which was done through a new equity issue without financing.

Kazuo Inamori accepted AVX's repeated requests for changes in the equity swap rate, in order to secure a smooth amalgamation while preventing any decline in AVX employees' morale. Thanks to such a strategy, AVX avoided the earnings decline that usually accompanies any acquisition. It has grown fast as a member of the Kyocera group.

AVX shareholders benefited greatly from Kyocera's AVX acquisition, and the AVX share price rose sharply on the New York Stock Exchange. AVX, as a Kyocera subsidiary, was later again listed on the New York Stock Exchange, earning Kyocera a huge profit.

Enhancing Information Technologies — Omron

Omron acquired Nippon Data General Corp. in 1991 and renamed it Omron Alphatec Corp. in 1996. Nippon Data General originated from Nippon Mini Computer Corp. founded in 1971. Nippon Mini Computer was renamed Nippon Data General in 1980 when it received investment from Data General Corp. of the United States. Serving as the Asian base of the U.S. firm, Nippon Data General saw its earnings deteriorate with the emergence of open-system computers in the late 1980s.

Nippon Data General was acquired by Omron under such circumstances and remained in deficit for five years after the acquisition. But its earnings recovered with the growth of its client server business. Nippon Data General, renamed Omron Alphatec, has developed into a core component of the Omron group's information technology business.

The abundant human resources at Nippon Data General have been very valuable for Omron. While eliminating gaps between computer manufacturers, the emergence of open-system computers prompted computer marketers to have excellent system-building capacity. For Omron Alphatec, the large number of excellent engineers who had been a major cost factor has brought about a competitive advantage.

Upgrading Core Competences — Murata, Rohm, Horiba

Murata bought out its U.S. rival, electronic component manufacturer Erie Technological Products Ltd., in 1980. Earlier, Murata's goal was to become a company like Erie, which featured excellent technological levels and was a forerunner in ceramic monolithic capacitors. The acquisition allowed Murata to advance into the industrial electronic machinery field and obtain U.S., Canadian, Mexican, and German production bases, and U.S., Italian, and French sales firms. The deal thus promoted the further globalization of Murata.

Rohm became the first Japanese firm to found a subsidiary in Silicon Valley in 1971. Through its IC manufacturing subsidiary, Exar Corp., Rohm acquired Excel in 1986. Exar made an initial public offering on the Nasdaq market in 1985. Rohm sold all Exar shares by 1994.

Excel was a Silicon Valley venture with basic patents covering flash memory. It was in dispute with Intel Corp., which had a great number of patents on flash memory applications. Excel eventually settled the dispute while protecting its basic patents.

Using the basic patents, Rohm in 1996 became the first company in the world to commercialize a nonvolatile ferroelectric memory, which was viewed as an ultimate memory indispensable for next-generation information terminals. Nonvolatile memory does not lose memory contents even if the power is turned off. A ferroelectric memory generates power with its coil upon receiving electric waves externally and does not have to be connected to any power source. The nonvolatile ferroelectric memory is useful for smaller-sized terminals since it neither has to be connected to any power source nor loses memory contents even with the power turned off.

While having such sophisticated technology, Rohm has refrained from producing any general-purpose components competing with the products of large computer chip manufacturers. It exploited cutting-edge technologies obtained through corporate acquisitions to enhance its core competence. Rohm has maintained its competitive advantage by focusing its business resources on custom-made components, instead of undertaking mass memory production, which would incur huge investment costs.

In 1996, Horiba acquired ABX S.A., a French blood cell counter manufacturer with which the firm had maintained technical and sales cooperation. In making the deal, Horiba paid attention to ABX's excellent technical and management capabilities.

Blood cell counters can provide massive data about human health conditions. By taking advantage of the ABX acquisition, Horiba obtained the blood cell counter technology and enhanced its medical device division. Furthermore, the acquisition allowed Horiba to have a blood cell counter production base and expand its sales arm to sell the product in China and Southeast Asia as well as Japan.

Chapter 4

Corporate Philosophies Emphasized in Management

Kyoto High-Tech Companies Emphasize Corporate Philosophies

The corporate philosophies here, as mentioned above, cover corporate values as materialized by mottos, mission statements, business management philosophies, management beliefs, or the like.

A corporate philosophy has a macro aspect that gives employees specific guidelines their company follows, as well as a micro aspect that provides employees with standards for their daily business decision-making. The corporate philosophy is the foundation involving the whole range of corporate management including management decisions, organization, production, R&D, and marketing. Unless a corporate philosophy is firmly established, business managers may lose touch with themselves in a global market where speedy actions are required. Unless a corporate philosophy is well understood by employees, their company's decision-making standards may be left unclear for employees. This may cause them to make the wrong decisions.

A corporate philosophy directly expresses a top leader's management thought and defines the organizational characteristics of a company. Organizational characteristics create unique organizational capabilities and have an impact on the creation and implementation of strategies (Okumura, *Japan's Top Management*, p. 156). In this sense, corporate philosophies are very important for business management.

Management leaders of the Kyoto high-tech companies have sufficiently understood the importance of corporate philosophies and made persistent efforts to communicate them to employees.

Kazuo Inamori says:

If an organization is to function well and make achievements, its goal should be specified and all organization members' vectors should be adapted to the goal. In an enterprise, a code of conduct called a corporate philosophy or creed may work to adapt employees' vectors to its goal. The code must be based on a fundamental thought or philosophy. Since the days immediately after Kyocera's founding, I have compiled what I have learnt while striving to live better as the "Kyocera Philosophy" and have tried to share the philosophy with all employees (Inamori, Respect the Divine and Love People: What Have Supported My Management, p. 22).

Such an unrefined management posture might have worked to boost employee morale and make great achievements.

As indicated in Table 6, the corporate philosophies of the Kyoto high-tech companies are very unique. Their key points are that maximization of corporate earnings should be viewed only as one aspect of corporate operations; also, employees working for a company should be happy, a company should be accountable to its shareholders, and a company should make a contribution to the local community.

The motive for working includes not only salary but also the willingness to make good products and provide good services, or to do a valuable job. The pleasure of making a contribution to society through doing a valuable job can lead people to do a good job.

The corporate philosophies of the Kyoto high-tech companies indicate a full understanding of this point and directly express key points of corporate operations.

Omron founder Kazuma Tateisi cited the following eight conditions that should be developed for the growth of a company (Abe, *A Study on Kazuma Tateisi*, p. 186):

1. Corporate philosophy
2. Instinctive behavior

Table 6. Corporate philosophies of Kyoto high-tech companies

	Kyocera	Omron	Murata	Rohm	Horiba
Motto	Respect the divine and love people (preserve the spirit to work fairly and honorably, respect the divine, and love people, our work, our company, and our country)	(Company motto) At work for a better life, a better world for all	For the purpose of our company's development and coprosperity with our partners, we shall refine technologies, practice scientific management, and provide unique products to contribute to the advancement of cultures; along with people who are pleased with and thankful for our company's development and coprosperity with partners, we shall manage our company	(Company mission) Quality is our top priority at all times. Our objective is to contribute to the advancement and progress of our culture through a consistent supply, under all circumstances, of high quality products in large volumes to the global market.	Joy and fun
Corporate philosophy	To provide opportunities for the material and intellectual growth of all our employees, and contribute to the advancement of humankind and society	Offer maximum satisfaction to customers; adopt a challenging spirit; focus on gaining our shareholders' trust; respect individuals; become a responsible corporate citizen; maintain corporate ethics while promoting corporate activities	Do not imitate others or be imitated by others		We shall grow endlessly toward a more bountiful future

3. Profit sharing
4. Job satisfaction
5. Participation (principle of self-containment)
6. Putting a company in a growing market (new market exploration)
7. Independent technology
8. Leadership

Kazuma Tateisi has emphasized that the corporate philosophy comes first and is the most important.

We here would like to look into the Kyoto high-tech company founders' thoughts and philosophies as well as the background for their development of corporate philosophies. We would also like to review how these companies' managers have tried to communicate these corporate philosophies to employees.

How Have Corporate Philosophies Been Developed?

The founders of the Kyoto high-tech companies have given interesting stories on how they have developed their corporate philosophies.

Kazuo Inamori said of his corporate philosophy:

> *I found that business management is "not the realization of my dream but the protection of employees' and their families' lives now and in the future."*
>
> *From this finding, I learned lessons that any manager should do their best to make employees happy and that any company should have legitimate reasons apart from any manager's personal interests.*
>
> *Then, I put "to provide opportunities for the material and intellectual growth of all our employees" at the top of the corporate philosophy and added "to contribute to the advancement of humankind and society" in order to fulfill the responsibility as a member of society. I made the two points to develop Kyocera's corporate philosophy. (Inamori, Respect the Divine and Love People: What Have Supported My Management, p. 43)*

As indicated by the corporate philosophy, Kyocera based its management on a partnership between its employees and encouraged all of them to participate in management.

Kyocera's motto, "Respect the Divine and Love People," means protecting and respecting the divine as the reason, and loving fellows humanely. The motto is described as to "preserve the spirit to work fairly and honorably, respect the divine, and love people, our work, our company, and our country."

Inamori has always told employees, "We should be the servants of our customers." This represents an ultimate customer-oriented business strategy that gives top priority to customers' requests in all areas from research and development to marketing. The strategy can work only because the motto has been communicated to employees.

Omron founder Kazuma Tateisi gave the following explanation about the company motto, "At work for a better life, a better world for all":

> *Our daily work first grows the company. Our belief is that the company has no choice but to be grown. The motto means that we should grow our company in order to make a greater contribution to society (Abe, A Study on Kazuma Tateisi, p. 104).*

By emphasizing the social presence of the company to demonstrate that employees' work would bring about social advancement, the motto serves to boost their business morale.

Omron has the special declarations — Corporate Citizenship Declarations, Environmental Declaration, and Corporate Ethics Declaration — as supplements to the concept word or motto as given in Table 6. Omron's corporate philosophy is illustrated in Figure 14.

It is natural for Omron as a high-tech manufacturer to boost customer satisfaction for its growth. At the same time, however, Omron emphasizes its consideration for shareholders, local communities, and the global environment. Such a corporate philosophy has become Omron's spiritual backbone.

Around 1953, a major radio receiver manufacturer asked Murata to downsize capacitors. However, as Murata was then playing a key

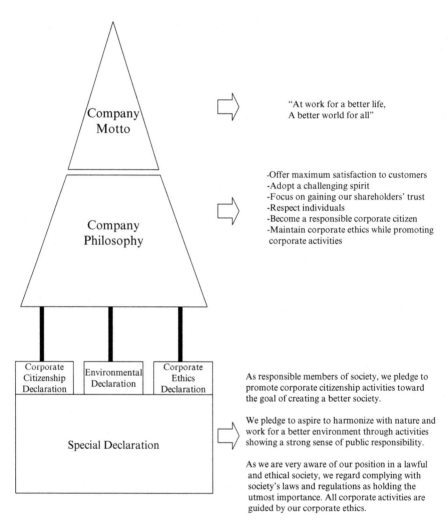

Fig. 14.   Omron's corporate philosophy (From Omron homepage)

role in developing Japanese Industrial Standards and Defense Agency Standards and was busy with standardizing components, it turned down the request from the radio receiver manufacturer. As a result, capacitor orders shifted to other companies. This misjudgment, coupled with the subsequent general economic deterioration, plunged Murata into a financial crisis and forced them to solicit the voluntary retirement of some employees.

Learning lessons from this mismanagement, Chairman Akira Murata developed a corporate motto specifying a basic guideline for business management to avoid any crisis leading to the solicitation of employees' voluntary retirement. The motto says: "For the purpose of our company's development and co-prosperity with our partners, we shall refine technologies, practice scientific management and provide unique products to contribute to the advancement of culture. Along with people who are pleased with, and thankful for, our company's development and co-prosperity with partners, we shall manage our company."

Since then, Murata has implemented scientific management including the development of step-by-step cost control, equipment investment accounting systems, and equipment productivity and other numerical indicators. It has also developed black-box technologies and modular products that cannot be imitated by other companies.

Rohm's corporate philosophy (company mission) says: "Quality is our top priority at all times. Our objective is to contribute to the advancement and progress of our culture through a consistent supply, under all circumstances, of high quality products in large volumes to the global market." This philosophy is closely related to the origin of the corporate name of Rohm. The name combines R for resistor with "ohm" as the electric resistance unit. The resistor products, which led to the founding of the company, were thus adopted for the corporate name. The letter R coming at the top of the name also means "reliability" reflecting the company mission's top sentence: "Quality is our top priority."

Horiba's corporate philosophy is "fairness." The philosophy was established as its founder Masao Horiba tried to disclose financial information and distribute profits fairly to J. Osawa & Co. Chairman Yoshio Osawa, Keifuku Electric Railroad Co. president Yoshijiro Ishikawa, and others who made a personal investment in Horiba's founding.

Masao Horiba related an episode where he had adopted the unique corporate motto of "Joy and Fun":

> *Doing something positively is far different from doing the same thing reluctantly. When we believe that doing something is interesting, we can enjoy doing it. Life should be joyful and funny. A motto of mine should fit the company. I courageously proposed the motto of "Joy and Fun" at a board meeting. (Horiba, My History, Nihon Keizai Shimbun.)*

As a matter of fact, many participants in the board meeting raised objections to the proposal. But Horiba successfully persuaded the board to accept the motto in commemoration of his shift to the chairmanship.

Above all, the founders of the Kyoto high-tech companies have developed corporate philosophies or mottos that express the founders' personal philosophies they have learned from their personal experiences and viewed as the most important. They aim to harmonize a company's pursuit of profit with employee happiness and the company's social presence.

Communication of Corporate Philosophy

Each of the Kyoto high-tech companies has made persistent efforts to communicate its corporate philosophy to its employees. Unless each employee understands the corporate philosophy, a company's decision-making standards may be left vague and affect the criteria and accuracy of decisions. In daily work, employees must base their decisions on the corporate philosophy.

The secret behind the strength of the Kyoto high-tech companies is that corporate philosophies are so thoroughly communicated to employees that employees as well as managers keep the core values of their respective companies in mind and can make decisions based on these philosophies.

Kyocera has continued a training program to give all employees lectures on its corporate philosophy since 1994. The program is not part of any in-house training curricula but an independent one for the employees to learn only the philosophy. Under the program, lecturers, including board members, explain about the "Kyocera Philosophy," a collection of Kazuo Inamori's messages.

Kyocera President Kensuke Ito says:

[The Kyocera president] is not only responsible for the company but also required to play a role in spreading the philosophy among employees. Spreading the philosophy is the most important job of the president. (Nihon Keizai Shimbun, June 22, 1998)

Unless employees unify basic values, their company may lose its fighting power. (Nihon Keizai Shimbun, June 19, 1996)

These remarks indicate that he understands the important impact of the corporate philosophy including personal thoughts.

Omron President Yoshio Tateisi used to hold 30-min "Morning Wind" meetings before the start of the work day to frankly discuss the corporate philosophy and other topics with young employees as far as his schedule allowed him to do so. The meetings were designed for President Yoshio Tateisi to give young employees brief comments on the corporate philosophy and the like.

Currently, Omron regularly holds "The Kurumaza (ring)" and "The Hizazume (face-to-face)" meetings. The Kurumaza, similar to the "Morning Wind" meeting, is designed for the president to have frank discussions with young employees while touring business offices throughout Japan. The Hizazume is a meeting for his communication with board members.

At Murata, all employees chant its corporate motto every morning. Its in-house workshops include explanations about the motto. Leaflets that are a compilation of the business management philosophy of founder Akira Murata have also been distributed to employees.

The top managers of the Kyoto high-tech companies have a sense of crisis regarding the loss of corporate philosophies. They have continued to seize every opportunity to communicate corporate philosophies like mottos and management principles to their employees.

Chapter 5

Unique Organizational Management

Until the late 1970s, the Japanese market had featured overdemand, where demand exceeded supply. It had then been important to produce good products just in time. In the era of oversupply since the 1980s, when supply has exceeded demand, it has been important to positively look for and find market needs, promptly develop products, and grasp business opportunities. In this sense, top corporate managers must grant considerable authority to employees and collect opinions promptly from frontline employees struggling in the marketplace for timely decision-making.

In the semiconductor and other industries that feature fast-changing markets and technologies, keen market competition, and many uncertain factors, economic efficiency generally increases as decision-making authority is delegated to lower levels within systems, and procedures such as decentralization are promoted (Galbraith and Nathanson, *Strategy Implementation: The Role of Structure and Process*, p. 63). Decentralization allows organization units to respond to environmental changes autonomously. At the same time, it encourages frontline units to activate learning and accumulate knowledge and know-how.

An organization consisting of small decentralized units has been the key to the Kyoto high-tech and other companies that have sustained top market shares in fiercely competitive global markets. They have tried to take advantage of decentralization to draw strength from group dynamics.

Visionary companies like 3M and Hewlett-Packard have also promoted decentralization and delegation of authority to lower levels, developing organizations that can solve problems flexibly. The Kyoto high-tech companies have achieved high growth with similarly unique organizational management.

The Kyoto high-tech companies have also strictly controlled costs to boost profits. To this end, they have devised ingenious organizational management methods. They have reduced the size of organizational units to control in a bid to thoroughly promote cost savings. Specifically, each unit uses unique management indexes to control costs and produces monthly earnings statements.

Exploitation of Information Technology

The Kyoto high-tech companies have paid attention to and concentrated their business resources on their respective core competences, developing management systems that can maximize gains. Such systems include Kyocera's ameba management, Omron's producer system, and Murata's matrix management.

These management systems are similar to the network organization, called the Business Organization of the 21st Century, as the business organization structure has shifted from the traditional functional organization to the divisional or matrix organization. The network organization is a new organization model, first seen around 1990, in the history of the transformation of the organization structure, as illustrated in Figure 15. In the network organization, the organization structure and form can be changed flexibly to meet market needs for speedier R&D, production, and other business operations.

Information technology plays a key role in forming a network organization. Only IT development can allow a network organization to be formed. More specifically, the IT development can activate horizontal and vertical communications to allow organizational components to share information (knowledge) while bringing about faster decision-making by business managers. IT development can also enable a business organization to accumulate knowledge. This

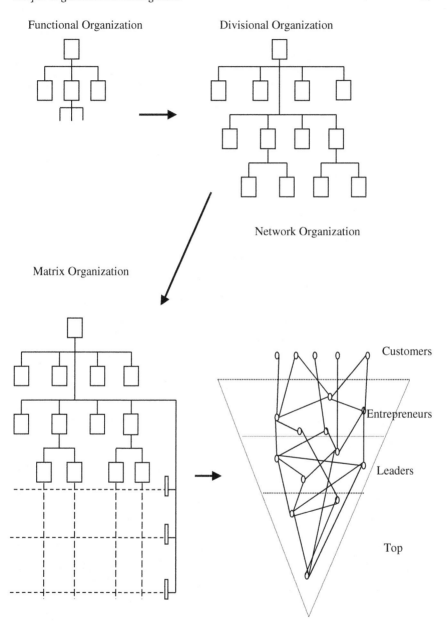

Fig. 15.   Transition of corporate organization

can lead to the triggering of innovation for individuals and their company.

IT plays a key role in business management. Although cost control at levels of networked small organizations and individuals may seem to be impossible, IT development has made such control possible.

The Kyoto high-tech companies have fully utilized IT to promote information-sharing communication and control costs. With this point in mind, we would like to analyze the Kyoto high-tech companies' organizational management, which has brought about robust earnings and growth.

Kyocera: Ameba Management

Kazuo Inamori said of profit and costs:

> I have believed: "Profit is no more than a gap between sales and costs. If so, we should make efforts to maximize sales and minimize costs. Profits should emerge as a result of such efforts." Since its founding, Kyocera has tried to develop the organization and management systems that are required for maximizing sales and minimizing costs. (Inamori, Respect the Divine and Love People: What Have Supported My Management, p. 52.)

Dissatisfied with the ratio of pretax profit to sales and other ordinary measures of profitability, Kyocera has achieved robust sales and profit by seeking to maximize sales and minimize costs. Every employee has been trained to believe that costs should be saved even if sales are expanding.

Inamori devised the "ameba" management system for profit-maximizing. Amebas are small business units that can expand and shrink flexibly and autonomously as needed to eliminate business waste while allowing each employee to work vigorously.

Inamori said of the ameba management system:

> A larger organization has difficulties in identifying waste. As Kyocera grew its organization enlarged, I thought that a large

organization should be divided into small ones (cells) to avoid wastefulness. I also thought that small organizations should be designed to allow each employee to exercise his or her abilities to the maximum extent in order to work vigorously. From this viewpoint, I devised the ameba management system. (Inamori, Respect the Divine and Love People: What Have Supported My Management, p. 95.)

Amebas at Kyocera are small units that can expand and shrink elastically and flexibly as needed. Amebas are organized around products, processes, or clients without any rigid rules. They can expand on growth in work volume and shrink on decline. They are protean, changing at will. Each Kyocera division has been a framework for such amebas.

Kyocera considers various measures to reduce working hours and expand sales at the ameba level rather than the division level. Amebas are designed to serve as small in-house companies and trade with each other. Each ameba is a self-supporting unit. For each ameba, various business data are published, including daily output value, raw material costs, and depreciation of machine tools. Ameba leaders meet every month to report their respective business performances. Any ameba with inferior performance could be absorbed by, or merged with, other amebas, or divided further.

Each ameba has a single leader, with no other supervisors. Each ameba leader is responsible for the management of his or her ameba. All employees within each ameba cooperate in improving their ameba's business performance. In this way, all employees participate in the management of their respective amebas and the whole of the company.

Kyocera has developed per-hour earnings indicators since its introduction of the ameba management system. Each ameba is evaluated according to an independent profit system and each ameba member according to per-hour value added. This is called an hourly profitability system.

The hourly profitability system is specified as follows: (Inamori, *Kazuo Inamori's Practical Theory: Management and Accounting*, p. 125):

> Gross shipments (gross output) − In-house transactions
>   = Output (net output)
>
> Output − Deductions (costs for raw material purchases, outsourcing, and personnel) = Net sales (value added)
>
> Net sales/Total working hours = Net sales per hour

The hourly profitability system has been developed to introduce business indicators that any employee can easily understand, even without any financial or accounting knowledge. Anyone can understand these indicators without specialist knowledge. Net sales are a concept close to value added and used for improving the value-added productivity.

In 1993, Kyocera also introduced hourly sales. The indicator serves to require each ameba to rationalize business management by increasing sales without any rise in working hours or by shortening working hours even in the absence of sales growth. Hourly sales are an indicator that is more understandable than hourly profit (*Nikkei Sangyo Shimbun*, November 11, 1996).

Each ameba in the marketing division divides sales in a month by total working hours in the month to get sales per employee. Similarly, each ameba in the manufacturing division divides production value in a month by total working hours in the month to get production value per employee.

If an ameba were to expand hourly sales, it would expand sales (production value for an ameba in the manufacturing division) or reduce working hours in response to flat or falling sales.

A key point of the ameba management system is that salaries are not directly linked to each employee's hourly profitability (or value added) or sales. Kyocera does not provide any special bonus or other financial incentives to reward an employee for his or her conspicuous contributions to the company.

The system is designed to lead every employee to understand business indicators and become conscious of participation in business management. It is for all employees to take part in the management of the company. Under the ameba management system, amebas can receive a spiritual honor for making excellent achievements and their members become more conscious of participation in such achievements. This can serve as a driving force to boost the morale of the whole of the organization.

Supporting the ameba management system is Kyocera's information system.

Kyocera has developed a management accounting system to prepare daily data on details of sales, output, costs, and in-house transactions, allowing every employee every day to see each division's profitability for the previous day. The system provides output, sales, procurement costs, distribution costs, and other data for each ameba, allowing a profitability management sheet to be developed. Such data for one day are published throughout the company the next day to become available for top managers and ameba leaders to make daily decisions.

In the course of past system development, Kyocera has worked out the "principles for system development," as illustrated by Figure 16, which will be applied to future system development.

Currently, the company is developing the "Kyocera information highway" as its comprehensive information network, which is planned to replace the present management accounting system. This will allow managers and employees to get real-time business management indicators on a global basis, and serve as the information-sharing infrastructure.

Omron: From Producer System to Holding Company

Since its founding, Omron has positively promoted the decentralization of management and the transfer of authority to capable personnel to thoroughly manage earnings and speed up decision-making. This has amounted to the implementation of Omron founder Kazuma

Kyocera's Principal for System Development

Principles   Objective

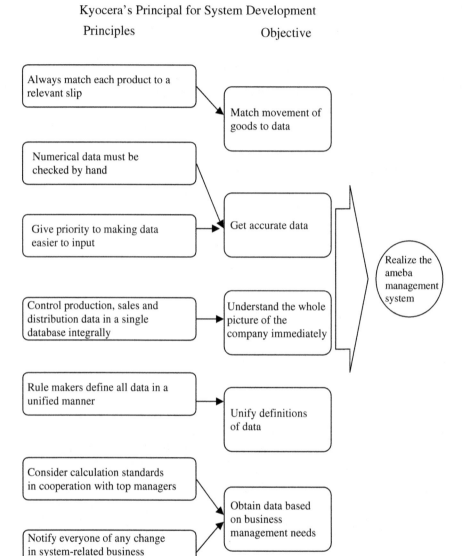

Fig. 16.   Kyocera's principles for system development (From *Nikkei Computer*, October 14, 1996, p. 132)

Tateisi's "Social Needs Theory." This theory represents a business strategy where Omron anticipates what society needs as early and as deeply as possible and develops technologies, products, and systems to meet those needs.

If the Social Needs Theory were to be put into practice, any strategy would have to be worked out in accordance with market conditions. Since around 1955, Omron has developed a system for the limited production of diversified products to flexibly meet market needs. Even now, the limited production of diversified products is difficult for many companies. Omron's system for producing a variety of products in small lots has been the so-called Producer System.

The Producer System is for Omron to establish wholly owned subsidiaries that are designed to pursue their respective independent profitability. Omron takes intensive control of their order receipts, technology and product development, financing and personnel recruitment. Omron founded many small subsidiaries, each for small-lot production, allowing the Omron group to produce a great variety of products.

The Producer System for the small-lot production of diversified products seems to be inefficient for the Omron group as a whole. But each subsidiary can produce a limited range of products each in large lots. Eventually, the system allowed Omron to achieve very efficient and rational management. In this sense, the Producer System may be viewed as a combination of centralization and independent profitability. This is an excellent organization management method for the decentralization of management.

However, the decentralization-oriented organizational management grew rigid. Around 1983, Kazuma Tateisi described Omron's condition as the "big-company disease." Its organization was then so complicated that front-line information took too much time to reach top managers. This led Omron to lag behind in developing products and responding to market needs. The result was a loss of customers.

Kazuma Tateisi then thought the big-company disease would be cured if Omron went back to the original concept of the Producer

System to create a large number of small appropriately-sized divisions that would be given as much authority as possible and managed as in-house small companies. He tried to go back to the venture business management as during Omron's start-up period.

In 1982, Omron reformed its organization and came up with the following venture management measures:

- Dividing strategic business units into smaller groups to enhance decentralization.
- Expanding each SBU leader's supervision and control authority.

Each SBU, corresponding to a division, was empowered to handle everything from research and development, production and domestic marketing to overseas marketing.

This management method has been maintained in an evolved form. Earnings, which had been controlled by divisions, are now controlled by smaller product groups. Each product group prepares a profit and loss statement to enhance control of earnings. There are some 100 product groups. For each product group, Omron subtracts material, labor, and other sales costs from sales to get profit.

Each product group has a project manager who is in charge of controlling earnings, development time for new products, and delivery schedules for each production step. Each project manager is empowered to make decisions for any part of the whole process from product development to termination of production. As a result, speedier decisions are made on each product.

The project manager system serves to shorten the distance between the marketplace and the company, allowing decisions to be made at levels that are closer to customers. The transfer of authority to project managers is working to further flatten the Omron organization.

In 1997, Omron introduced an internal company system, giving senior general managers the authority to plan business strategies and make equipment investment decisions for their respective divisions. Under the system, return on assets and other financial indicators are used to manage earnings for each division or internal company.

Omron is positively considering introducing a holding company system if certain conditions are met.

Incumbent Omron president Yoshio Tateisi said of the future roles of the president, directors, and middle managers:

> *What the president will do is to specify the company's vision and ideals and develop the conditions necessary for their realization. The president will be the company symbol that all employees will trust. Each director will exploit information and network technologies to control a workforce that will be three to five times as large as the present size. The president will give each director the authority to play a more important role in making decisions ... Middle managers' roles will shift from the conveyance of their seniors' instructions to training and taking care of employees. They will be granted more authority. (Nikkei Sangyo Shimbun, November 8, 1996.)*

President Yoshio Tateisi defines the roles of the president, directors, and middle managers in this way. Especially, he defines middle managers as "managers closest to the marketplace." Using the word of "middle manager-oriented management," he promotes their positive participation in management. This may amount to the implementation of Omron founder Kazuma Tateisi's Social Needs Theory.

Omron has positively exploited information technology to support such earnings management and to flatten its organization. Ahead of other Japanese companies, Omron provided all managers with personal computers. It has aggressively introduced groupware and intranet systems to promote the sharing of information and faster decision-making.

Currently, Omron is developing an electronic collective decision-making system, a marketing database, and a production control network. It has completed a system that allows the company to see each group's earnings on a global consolidated basis every month. Omron group companies, both in Japan and overseas, provide monthly earnings data to the system that enables the head office to manage earnings data integrally.

Murata: Matrix Management of Smaller Units to Control

Under the motto of scientific management, the Murata group has introduced the matrix management system since around 1960 to secure independent profitability for each product and each process. Under the system, Murata has subdivided its organization into the smallest possible units for better cost control. As a result, Murata is now able to make a profit even if the capacity utilization rate is as low as 60%.

The matrix management system originated from processes for the production of ceramics, as Murata's mainstay product. It has been devised to control costs more accurately by allowing Murata's raw materials for ceramics to be compared with those of other specialized ceramics manufacturers. Ceramics production processes include the mixing of raw materials, the baking of ceramics, and the processing of baked ceramics. Murata calculates the costs for each process under the assumption that transactions have been made between processes. The process-by-process cost calculation allows the company to pinpoint any problem areas, even for a product that takes a long time to complete.

This process-by-process control has developed into the matrix management system to control costs for each division, as well as for each product, including capacitors, filters, and electronic devices. The matrix management system is designed not only to minimize costs but also to lead each process or division to increase its management efficiency. Specifically, interest rates are levied on the assets used by each process or division, prompting each process or division to pay attention to its asset turnover ratio.

Under the matrix management system, each process is managed as an independently paying enterprise to maximize its profit. The system's ultimate objective is to maximize value added for the whole of the Murata group. The organization is subdivided into the smallest possible units that are easy to control and are responsible for producing profit independently. The concept of the matrix management system is illustrated in Figure 17.

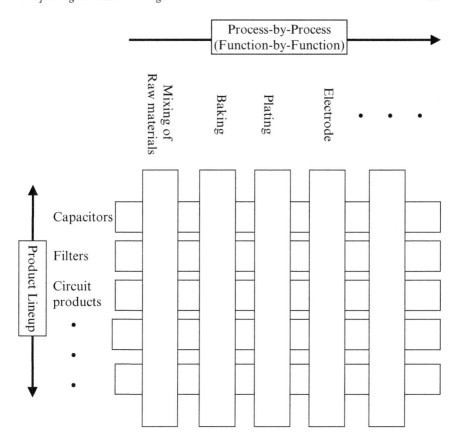

Fig. 17. Murata's matrix management

Each product- or process-based matrix controls its profit and loss, allowing the Murata group as a whole to increase the efficiency of cost control and management. The number of such independently managed matrixes in the Murata group totaled 2,500 in Japan alone. As the Murata group has subdivided its organization into the smallest possible management and control units under the matrix management system, it has become sensitive to profitability, prevented the parent from becoming bloated and been ready to respond promptly to organizational reforms including the spinning off of some units.

Information technology has been supporting the matrix management system. Murata has developed a system to see a day's sales in the afternoon of the next day. The daily sales data are broken down by business or product category. Daily sales are specified for each of the overseas and domestic marketing divisions, products, and product groups. This system has played a key role in backing up the matrix management. Murata has also introduced an information system to control progress in the production and quality of products, as well as a database system for the sharing of marketing and technology information throughout the Murata group.

Since 1996, Murata has introduced an electronic collective decision-making system to speed up decision-making and transfer authority to lower levels. A key point of Murata's electronic collective decision-making system is that decisions are published through the intranet within the company as soon as they are made. This allows all Murata employees to share information and check decisions. Murata is thus able to check any abuses and other problems that can accompany a transfer of authority.

Rohm: Horizontal Organization of Small Units

Rohm has exploited outmoded technologies to steadily produce custom products that meet specific orders from customers, including electrical machinery manufacturers, while ignoring cutting-edge products like dynamic random access memories that large semiconductor manufacturers have been racing to develop with all their technological resources.

This demonstrates that Rohm has been well aware of its core competence. Rohm has refrained from undertaking product development where massive financial and human resources are required, in consideration of its financial and human capacity. However, it has achieved robust profits by focusing its business resources on custom products that are not produced by other companies but lead to stable profits. Other companies have been reluctant to manufacture custom products that take a lot of time to develop. Rohm has also refrained from handling finished products whose production and sales require a large amount of capital and after-sales services. Viewing itself as a manufacturer of components, Rohm has focused on components rather than on finished products.

Rohm's production setup is also designed to save costs. Products whose costs have soared are transferred to domestic and overseas subsidiaries along with the production technologies. These subsidiaries have originated from plants that Rohm had commissioned to manufacture products when it was still a small start-up company. Rohm has thus commissioned subsidiaries to manufacture products that it cannot produce. It has thus initiated the present form of a fabless company. As a result, Rohm has naturally developed as a flat group of small companies.

Rohm has a unique system for the appointment of managers. Employees' promotion to managerial posts must be recommended by their direct bosses first. Candidates for middle managers must be interviewed by all the directors. These interviews are designed to select only qualified employees as middle managers. Rohm does not necessarily fill all managerial posts. If no qualified employees exist, managerial posts may be left vacant.

Rohm has also introduced information technology aggressively. It has introduced a monthly reporting system to collect all in-house data every month. Monthly reports take the form of account settlement tables based on corporate accounting principles. Each of the manufacturing and marketing divisions prepares a monthly report, preventing any vague report from being produced. On a monthly basis, Rohm discloses sales, cost ratios, operating profit, pretax profit, net profit, and other data for each division and product. Its monthly report ranks business units by an earnings indicator, allowing each unit to know its ranking within the company.

Rohm has an information network linking overseas subsidiaries and LSI centers at home and abroad to allow advanced exchange of information and faster decision-making. It has also developed a company-wide technology database.

Horiba: Product-Oriented Management System

Horiba has adopted a product-oriented management system where the organization is divided into groups that cover research and development, production and marketing for their respective product categories. Each group specializes in a specific product category, integrating operations from development to sales for the category. This setup is designed to efficiently develop and produce market-oriented products for their timely provision to the market. Specifically, the organization is divided into four groups based on four product categories — engine measurement instruments, electronics and information equipment, environmental and medical equipment, and scientific measurement instruments.

Previously, Horiba had adopted a functional organization divided by function into research and development, production and marketing groups. The functional organization, though generally conceived as efficient, resulted in insufficient communications between the R&D and marketing groups. While the R&D group spent time for the development of promising products, the marketing group lost opportunities to successfully introduce the products into the market. The marketing group separated from the development and production groups tended to sell products that could sell well and refrain from pushing products that did not sell well then but could have become popular in the future.

The product-oriented management system can shorten the period of time between the development of products and their offer to customers. It is eventually more efficient than the functional organization. If a product group fails to make a profit, the group may try to find ways to cut production and distribution costs to become profitable and its marketing team may explore new markets. The group as a whole may try to devise various profit-generating innovations.

While giving priority to such a product-oriented vertical division of the organization, Horiba has horizontally introduced information and other platform technologies common to all product categories or markets. The vertical groups also outsource common operations to each other and some units are established to undertake common operations for all vertical groups horizontally. Horizontal units are introduced to complement vertical groups in order to improve efficiency. Horiba has thus developed a matrix organization of vertical and horizontal divisions.

Regarding information technology, Horiba has utilized electronic bulletin boards and e-mail to support speed-oriented management reforms including the "Time One Half" movement. It has also developed technology and marketing databases to allow employees to actively share technology and marketing data.

Chapter 6

R&D and Production Control, Cost Control and Personnel Management Systems

Introduction

As noted in Chapter 5, the Kyoto high-tech companies have successfully drawn on the strength of group dynamics by taking advantage of information technology and decentralized organizations consisting of small units to control. As well as small units to control, however, some more might have been required for these firms to practice product-oriented strategic measures, including prompt responses to market changes and large-scale innovations. In fact, it is significant to organically combine research and development, production, cost and personnel management systems to complement organizational management based on small units. Various systems must be linked to organizational management to successfully practice product-oriented strategic measures.

The Kyoto high-tech companies have developed specific management systems, as shown in Figure 10. We would like to analyze these systems.

Market-Oriented R&D

The ratio of research and development costs to sales stands at some 8% for Omron and Horiba and at 6% for Murata, indicating that they have given significant priority to R&D. But a high level of spending does not necessarily guarantee successful R&D. Unless products that the market needs are developed, R&D spending may achieve nothing. In this respect, the Kyoto high-tech companies have developed systems to efficiently develop market-oriented products that meet market needs.

At Kyocera, which has been good at developing technologies, researchers and engineers not only exchange information with marketers but also visit customers' offices to understand market needs. Their ultimate objective is to develop market-oriented products. Kyocera thus views R&D as part of the marketing effort, giving top priority to customers' needs in all areas of R&D, production and sales operations. Production and sales staffs are not separated. Even researchers and production engineers visit customers' offices to understand market needs as required. R&D operations are thoroughly oriented toward customers. This is the secret that allows Kyocera to efficiently develop products that are well oriented toward the market and can sell well.

Rohm has a similar system. A requirement for developing custom ICs, Rohm's mainstay product, is the capacity to design products meeting customers' exact orders, rather than cutting-edge processing technologies. In order to develop custom ICs successfully, one must communicate closely with customers to accurately understand their needs. In this respect, Rohm assigns engineers to specific products proposed by customers so that the most suitable components for customers' products can be developed. It has a flexibly adaptable development and service setup. Rohm engineers are allowed to take part in the customers' product-planning stage to provide efficient, effective product development services to customers and develop best-fitting components on their own.

Kyocera and Rohm have thus increased the efficiency of R&D operations through their engineers' marketing efforts including their direct contact with customers to collect information.

Murata has promoted in-house sharing of technology information to increase R&D efficiency. In a bid to practice scientific management as stated in its motto, Murata has developed the unique "strategic development process management" system to accurately boost engineers' morale. This is a management system to increase

the speed and success rate of R&D operations by selecting technologies among in-house data for groups that undertake specific
product development projects. To support the strategic development
process management system, Murata has encouraged employees to
share technology information by making use of the in-house technology database, in-house exchange of information through technology
forums and communications with European and American institutes.
Murata has thus taken full advantage of technology information sharing to speed up the development of products.

Horiba has stepped up the exchange of technologies through an
international network to secure its R&D efficiency. Omron has created an in-house venture system to make its R&D operations more
efficient.

In a bid to speed up the development of market-oriented products, Horiba has created a supreme R&D decision-making body
comprising the president and directors to focus on R&D operations
and has been proceeding with efficient development of products. In
1998, it introduced the just-in-time initiative in its R&D division,
in addition to its production division, to enhance control of schedules. The initiative is designed to allow products to be accurately
developed within short periods of time. In order to respond to cross-
border market needs regarding environmental and other problems,
Horiba has developed a worldwide R&D setup linking the three
poles of Japan, Europe, and the United States for the exchange of
technologies.

Omron has promoted R&D operations to devise "golden products." Golden products must meet three conditions — innovating
customers' lifestyles or production styles, keeping market leadership
through discrimination using technology accumulation, and increasing profit and value added. In order to produce golden products,
Omron has formed new business development centers and other new
organizations and created a golden technology system and a venture
capital system. The new business development centers are designed
primarily for young engineers' development of new businesses. The
golden technology system aims at development of products with

higher value added by giving certain leaders R&D authority and funds.

Focusing on Production Technologies

Rohm and Murata have successfully developed their competitive advantage by developing production technologies that others cannot imitate.

Rohm has been exploiting outmoded technologies for manufacturing custom ICs that meet customers' specific requests. Instead of using cutting-edge technologies for mass production to gain a competitive edge, Rohm has developed a flexibly adaptable marketing and production setup to keep a competitive advantage. It has developed and produced products using production technologies that cost less and are highly likely to be profitable.

In the semiconductor industry, only the top companies can make profits even by investing a huge amount of money in cutting-edge technologies. Rohm has focused on its ingenuity to develop profitable products even with outmoded technologies and production equipment. It is armed with excellent production technologies.

Rohm has traditionally developed production machines on its own. It established a machinery division for that purpose. One reason for the in-house development of production machines is that in-house development costs less than purchases from other companies. An even more important reason is that Rohm can accumulate machinery-manufacturing know-how to flexibly modify production equipment in response to future cost-saving requirements and changes in customers' product specifications. This is the way Rohm has enhanced its production technologies.

In an effort to increase productivity, Rohm has efficiently located production machines in its small head office semiconductor plant. This is because it can easily develop and customize production machines on its own. Rohm has also practiced mixed-flow production, using the same line for the production of different ICs and other products. This can accelerate depreciation of production machines.

Such a flexible production setup owes much to the firm's accumulation of know-how through the in-house development of production machines.

Rohm has assigned as many as 380 out of its 850 engineers to the production system group.

Under founder Akira Murata's philosophy that "new quality electronic equipment begins with new quality components that begin with new quality materials," Murata has grown by developing not only components but also their materials on its own.

Murata has a materials division that provides raw materials to its circuit products division. The materials division is designed to conduct research and development of qualitatively stable and more cost-competitive materials required for quality components and provide them within the company. Like Rohm, Murata has had a machinery division to manufacture major production equipment for in-house use since its founding. Just like in-house materials development, in-house manufacturing of production equipment has been aimed at stabilizing the quality of components. Currently, Murata is promoting the production of modularized products. In-house manufacturing of production equipment and modularized products is designed to prevent manufacturing secrets from being leaked to other companies. If a company purchases production equipment from the outside to manufacture products, other companies may buy the same equipment to produce the same products. A company can avoid such a situation by manufacturing production equipment on its own.

This approach is reflected in computer software development. Murata has a systems division that develops most of the in-house computer systems. In-house development of computer systems allows such systems to be customized more easily.

The vertical integration of technologies that Murata has achieved through long-term efforts for integrated production beginning with materials development has allowed the company to keep its technologies, manufacturing processes, and products in a black box. It has introduced production technologies that cannot be imitated by rivals, who reverse engineer Murata products to analyze technologies.

Coupled with the modularization of products, this has enabled Murata to keep the secrets of its products in a black box.

Murata has thus put into practice its motto: "Don't imitate others or be imitated by others."

Maximizing Employees' Value Added

The Kyoto high-tech companies have devised unique organization management systems to increase business efficiency and expand valued added by employees successfully.

Kyocera's ameba management system promotes the participation of all employees in company management to improve their morale for maximizing sales and implementing strict cost control for minimizing costs.

As a business management indicator, Kyocera has traditionally adopted the hourly profitability system, focusing on the expansion of value added based on each ameba's profitability. Hourly profitability is an easy-to-understand indicator covering only production value, costs, and total working hours of each ameba or group. This can clearly indicate how an ameba could increase its value added, prompting each employee to take part in company management and make efforts to increase value added.

In order to keep robust earnings in the electronic components industry, featuring a fierce price-cutting race, Murata has strictly controlled costs with its matrix management system to pursue independent profitability for each product or process.

The matrix management system prevents an organization from becoming bloated and leads management and control units to be minimized to strictly pursue profit. Product- and process-based management and control units are divided into the smallest possible units that can achieve independent profitability. Under matrix management, profit and cost is strictly controlled at each process. Consolidation of process-based profits emerges as a great achievement. As each process controls its profit just as an independent company does, small process-based profits can be accumulated as a huge value added.

Each employee works while considering his or her individual profitability. Each employee will then naturally make efforts to improve.

Murata has introduced productivity and capacity utilization indicators as well as the matrix management system. It has also adopted unique indicators such as floor area productivity and equipment productivity to help increase business management efficiency.

When Rohm closes its monthly books it provides employees with monthly earnings information in order to encourage employees, including managers, to remain sensitive to earnings. It has also made employees conscious of the importance of pricing by establishing a rule that any invoice prices agreed between the marketing and manufacturing divisions should never be changed.

Under this rule, the marketing division cannot change invoice prices even in the face of competition from rivals offering cheaper products. It has no choice but to step up marketing efforts to counter such competition. The manufacturing division, as well as the marketing division, is held responsible for the sales of products. This means that the manufacturing division is required to improve the quality of products to counter rival goods. In order to secure a profit, Rohm requires employees to decide whether new products have the required technologies and cost competitiveness to produce a profit when these products are introduced into the market.

Exercising Innovation

The Kyoto high-tech companies, characterized by the development of new technologies, give consideration to personnel management systems that allow employees to exercise their creativity. They have been focusing on developing self-starting employees who achieve innovations, rather than employees who only do routine jobs.

Rohm, Horiba, and Omron have introduced in-house prizes as nonwage incentives for employees.

Rohm has established five presidential prizes — the diamond prize worth 10 million yen, the gold prize worth 5 million yen, the silver prize worth 2 million yen, the bronze prize worth 600,000 yen, and the prize for efforts worth 50,000 yen. Middle managers and lower-ranked employees annually submit applications for prizes specifying their respective achievements, including development of products, cost savings, and other contributions to the company. An evaluation committee screens these applications to select prize winners. In fiscal 1997, prizes aggregated as much as 470 million yen for a total of 1,670 prize winners, including some 800 winners of bronze and higher prizes (*Weekly Toyo Keizai*, March 11, 1998, p. 136).

Horiba's in-house prizes are not as high in value as those at Rohm. Its in-house prizes are named after people who have contributed to the company since its founding. They are the Yoshio Osawa Prize for the marketing division, the Yoshijiro Ishikawa Prize for the management and control division, the Ko Takagi Prize for the financial division, and the Shinkichi Horiba Prize for the development division. They are awarded to employees who have made outstanding contributions to their respective divisions. In addition, a special overseas prize is given to those with distinguished achievements in the overseas division.

Yoshio Osawa, the second chairman of J. Osawa & Co., Yoshijiro Ishikawa, president of Keifuku Electric Railroad Co., and Ko Takagi, owner of a dyeing company, had been investors in Horiba upon its founding. Shinkichi Horiba, the father of founder Masao Horiba, was a professor at Kyoto University. These prizes also work to remind employees of their appreciation of these people and entrepreneurship.

Omron has an individual invention commendation system. It had originally established a commendation system for teams. Since fiscal 1996, it has introduced an individual commendation system to award up to 3 million yen to an individual for an excellent innovation. This is a management method to encourage employee innovation.

Horiba has four types of middle managers — the management manager for organizational and personnel management, the expert manager for administrative operations including accounting and marketing, the "meister" for machining skills, and the technical manager for electrical, mechanical, and other engineering fields. This is called the field-based manager system.

Among the four, only the managing manger is a real manager. The others are specialists in their respective fields and have no managerial role. The field-based manager system is designed to give suitable managerial posts to employees in accordance with their respective abilities and willingness, allowing them to fully exercise their skills. No wage gaps exist between the four types of middle managers.

Behind the introduction of the field-based manager system, the segmentation of the organization had expanded each middle manager's territory to unrealistically cover various fields ranging from marketing to research and development, production control and materials procurement. Any traditional concept of middle managers could not cover such a wide range of fields. At a time when the fast pace of technological progress leads to increased specialization in all fields, Horiba, whose core competence is research and development, must create globally competitive technologies and accumulate them as the company's knowledge in order to gain and keep a competitive edge. This background explains why Horiba introduced the technical manager and meister.

The field-based manager system is defined in Table 7.

A precondition for Horiba's development of the field-based manager system has been its unique merit-based employee evaluation system, which rewards employees on their respective merits. It is different from a demerit-based system, which gives a full score to each employee first and reduces the score in accordance with failures. The merit-based system simply awards credits in accordance with merits or achievements. This system encourages employees to fully exercise their abilities without the fear of failure. Morale can thus be boosted for the whole company. Employees can also be encouraged to voluntarily improve their skills as specialists in their respective fields.

In addition to these unique personnel management systems, Horiba has a limited four-day workweek system, designating the first Friday of every month as a holiday, and offers extra vacation for 10- and 20-year consecutive service to make the employees' jobs at the company more satisfying, or "joyful and funny," according to its motto.

Table 7.   Definition of Horiba's field-based manager system

Manager type	Definition
Management manager	The management manager as the head of a business unit is given authority and responsibility to achieve the unit's business targets, train deputies, make unit-level business plans, fix priority orders, share duties among deputies, coordinate with other units, and control progress in various aspects. The management manager serves as an interface between top managers and deputies
Expert manager	The expert manager exercises his or her excellent expertise and actions to solve problems and promote group efforts to achieve targets
Technical manager	The technical manager masters the world's No. 1 technology and expertise in a special field and becomes a model for "technology-oriented Horiba." In research and development operations, the technical manager exercises searching, research, planning, and decision-making abilities and tries to share technologies with others within the company
Meister	The meister masters the world's No. 1 skills in a special field, exercises production skills fully, and trains followers

(From: *Wage Practices*, March 1, 1994, p. 26)

Rohm has introduced an approach similar to Horiba's merit-based system, that is, a consolation-oriented personnel evaluation system. As indicated by the philosophy for Rohm's presidential prizes discussed above, this system is designed to re-evaluate employees based on annual achievements irrespective of their earlier performance. It can encourage employees to change their work attitudes any time even after failing to make outstanding achievements. Their morale can thus be boosted.

Just like Horiba, Murata has a unique personnel system to separate managers from specialists. In the past, Murata had based the promotion of employees only on management abilities and lacked any qualifications for specialists. But it has introduced qualification systems both for managers and for specialists to eventually improve the treatment of excellent specialist employees and boost specialists' morale. Managers on their part can focus on management as professionals.

Omron has made various efforts to make in-house organizations more entrepreneurial, including the establishment of a venture capital system as well as new business development centers for young engineers' new business development.

The new business development centers are directly controlled by the president and designed to search for new business opportunities in information and communications, traffic control system, healthcare, and other markets. These centers comprise primarily young engineers in their 30s, who are freely conducting research and development operations while being required to exploit their R&D achievements for business purposes. The problem here is how to link R&D achievements to sales.

Under the venture capital system, individuals or groups present their new technology or product proposals to a technology-oriented management policy council. After screening by the council, an outstanding proposal is given development funds totaling between 1 billion and 3 billion yen. A promising business proposal can be given additional funds and developed into a new business.

By the end of 1999, Omron created an unprecedented system to globally unify personnel evaluation systems. This system allows employees recruited overseas to compete fairly with those recruited by the head office in Japan, activating the head office's recruitment of foreigners.

Murata has unified qualification, wage, and other personnel systems for employees at both the parent company and the subsidiaries in Japan, practicing part of the system that Omron pursues.

Incentives for Learning

The Kyoto high-tech companies also encourage employees to obtain academic qualifications and attend in-house study meetings in order to upgrade their knowledge levels.

In a bid to increase incentives for R&D engineers, Murata allows these engineers as much as possible to publish their research through in-house magazines such as the "Technical Journal" and through academic forums. Research contributed to in-house magazines is put into a database available to all engineers. Presentation of studies at academic forums can develop business opportunities and pave the way for the engineers' acquisition of doctorates, serving as a strong incentive for engineers.

Horiba has traditionally been positive about the improvement of knowledge levels, as indicated by its founder Masao Horiba's acquisition of a doctorate in medicine.

At Rohm, lectures take place even on Saturdays and Sundays with outside lecturers invited. Lecture topics include LSIs and other technologies as well as marketing and planning. Employees participate in these lecture meetings voluntarily. If sufficient participants are enrolled in planned lecture meetings the company may provide the meeting venue as well as cover the relevant costs, including lecturers' fees. Omron allows its employees to select the lecturers on condition that they be leaders in their respective fields.

Chapter 7

Positive Social Contributions

Accountability

As indicated by their corporate philosophies, the Kyoto high-tech companies have done business in pursuit of their business activities' harmony with society and have also made various contributions to society. This is demonstrated by their emphasis on investor relations, including the disclosure of corporate information to investors.

Kyocera has been publishing not only consolidated and unconsolidated balance sheets and earnings reports but also sales and assets for each of its divisions including ceramics and related products, electronics, and optical and precision instruments, and for each of the regions such as Japan, the United States, Europe, and Southeast Asia. The firm has provided investors with these details through financial statements and the Internet. It has also published English language annual reports on the Internet.

Just like Kyocera, Murata has disclosed product-by-product and region-by-region business information in detail through the Internet. In addition to such segment-by-segment data, Murata has provided the ratio of operating profit to sales, the return on equity, and other business indicators for the past five years.

Omron and Rohm have provided English language annual reports for foreign investors, as well as Japanese language financial information, through the Internet.

Horiba has also been positive about disclosure of its information, providing investors and analysts with analyses of consolidated and unconsolidated financial statements for the past five years in the form of leaflets and CD-ROMs.

These analyses cover not only consolidated and unconsolidated financial statements but also the return on equity, the return on assets, and other profit indicators; the working capital ratio, short-term liquidity ratio, and other safety indicators; the total assets turnover, the ratio of personnel costs to sales, and other efficiency indicators; and stock prices and other investment information.

The CD-ROM takes advantage of video functions to introduce Horiba's profile and earnings, attracting not only investors and analysts but also ordinary citizens. The CD-ROM also includes tabulation files for investors and analysts to easily analyze earnings data.

The Kyoto high-tech companies have been very positive about the disclosure of financial information. Such an attitude has apparently contributed to their good reputation among overseas investors as well as Japanese, and to their high stock prices and the large number of their foreign shareholders.

Philanthropy

Since their founding as new ventures, the Kyoto high-tech companies have engaged in various philanthropic activities out of gratitude for the favors that society has given to them.

Kyocera founder Kazuo Inamori contributed his personal money to establishing the Inamori Foundation to launch the Kyoto Prize international awards in 1985. His personal contributions to the foundation have amounted to some 41 billion yen (20 billion yen in 1984 and 21 billion yen in 1997).

The Kyoto Prizes are awarded to individuals and groups for their significant contributions to the fields of advanced technology, basic sciences, and arts and philosophy. A Kyoto Prize laureate is selected irrespective of nationality, race, sex, age, or belief and awarded a diploma, a Kyoto Prize medal, and 50 million yen in prize money.

Kyoto Prize laureate candidates are screened strictly and fairly by the Kyoto Prize Committee and the Kyoto Prize Screening Committee in each field, and the Kyoto Prize Executive Committee.

Past Kyoto Prize laureates come from a wide range of countries, from the United States, Britain, Germany, and Japan to Poland, France, and Russia. Akira Kurosawa, a famed Japanese film director, was awarded the Kyoto Prize in arts and philosophy in 1994.

Rohm founder Kenichiro Sato, whose father was a violinist, began to learn to play the piano at the age of six, aiming to become a pianist. In his student days, he won the second prize in a music competition. But the failure to win the first prize prompted him to give up becoming a pianist. Sato then focused on manufacturing as his main interest and founded Rohm.

Such a background has kept Sato interested in music. He published *Visible Music History* in 1965, *History of Musical Instruments* in 1970, and *Asian Music* in 1981.

In 1991, Sato contributed some of his shares to launch the Rohm Music Foundation to support the musical arts. The foundation's objective is to support young musicians and artists, promote music culture through concerts, and contribute to the advancement of music culture through research and other academic activities. Regarding the training of young musicians, more than 25 musicians that the foundation has supported have won prizes in international and other competitions. Since 1993, the Rohm Music Foundation has sponsored the annual Kyoto International Music Students Festival. It has also tried to diffuse music culture by cosponsoring the Orchestre de Paris, the Prague Symphony Orchestra, and other concerts.

Murata established the Murata Science Foundation to subsidize and promote fundamental research, an area where Japan has lagged behind other industrialized nations. The foundation has sponsored research in natural sciences, including electronics, and human and social sciences such as economics, sociology, and international cultural issues. It has also provided assistance to academic societies and supported researchers' overseas study.

Omron calls its philanthropy "corporate citizenship activities" covering social welfare, culture and arts, science and technology and international exchange. On the anniversary of its foundation, May 10, called "Omron Day," the company conducts volunteer activities for the local community, including cleaning, tree planting, and visits to care houses.

Omron has also initiated social welfare activities to put into practice its corporate philosophy. Omron founder Kazuma Tateisi established Omron Taiyo Co. as a welfare plant for physically disabled people in association with Dr. Hiroshi Nakamura in Oita Prefecture in 1972. Omron Taiyo may be viewed as a plant only for welfare. But it is a business corporation pursuing profits.

Omron Taiyo assembled magnet relays upon its foundation and booked a profit of 1.8 million yen to pay a 10% dividend the next year. A key point is that physically disabled persons account for nearly 90% of Omron Taiyo employees, including the plant manager, and are given the opportunity to become directors.

The success of Omron Taiyo encouraged Omron to launch a second welfare plant in cooperation with Sony Corp. and Honda Motor Co. in 1981. Omron also established Denso Taiyo Co. in 1984 and Omron Kyoto Taiyo Co. in 1986. It has established plants one after another for the rehabilitation of physically disabled people.

Omron's cultural and artistic activities include lecture meetings such as the Omron Kyoto Culture Forum and the Omron Keihanna Culture Forum. Lecturers in various fields have been invited to give lectures at these meetings, which anyone can attend, free of charge.

Tackling Global Environmental Problems

The Kyoto high-tech companies have proactively tackled global environmental problems that have become a significant social issue expected to seriously affect future business activities. Their relevant efforts include the development of eco-friendly products, reduction of industrial wastes, recycling to save resources, energy-saving measures, and acquisition of the ISO (International Organization for

Standardization) 14001 Environmental Management Standard certi-
fication. The five firms' basic environmental policies are summarized
in Table 8.

Horiba has emphasized and positively tackled global environmental issues since its mainstay products have been analytical and
measuring systems including automobile emission analyzers. It has
developed and diffused analytical and measuring technologies and
products that are necessary for prevention of pollution and conservation of the global environment. The company has also provided the
public with information on global warming and other environmental problems. Since 1992, Horiba has sponsored the Honest Club,
a personal computer network for volunteers to provide information
on acid rain. At present, acid rain observation data from network

Table 8. Kyoto high-tech companies' policies on environmental issues

Corporate philosophy	Policy
Kyocera	Live together with the Earth
Omron	We will pursue the harmony between the environment and humankind and contribute to realizing a better environment through business activities, giving priority to the public
Murata	We shall look straight at the (environmental) issues anew and unite the whole of our organization to accumulate efforts to reduce environmental loads in a new sense and pursue the junction between environmental conservation and promotion of business efficiency
Rohm	Being fully aware that environmental conservation is essential to the continued existence of humankind and the permanent prosperity of enterprises, the Rohm Group will give full consideration to the conservation of the global environment in doing business
Horiba	Realizing that global environment conservation is the top priority for sustaining the affluent growth of humankind, we will work to challenge the limits of technology to ensure the harmonious coexistence of humans and nature and will contribute to creating an environment friendly to the Earth

members throughout Japan are made available on the Internet on a real-time basis.

Rohm has published numerical data and figures regarding its recycling of wastes for the past six years on its homepage. It has also disclosed energy-saving efforts regarding electricity and gas. Rohm has expanded its scope of accountability voluntarily to fulfill its responsibility to the local community by publishing data about the effects of corporate activities on the global environment, as well as financial information.

New Venture Development

The Kyoto high-tech companies started up as new ventures in the historic city of Kyoto. Therefore, their top managers have realized that it is important for them to adhere to the city and keep their entrepreneurial spirit.

On the other hand, a growing number of other companies are relocating their production base from Kyoto due to the increasing urbanization and tougher regulations for conservation of the ancient scenery. Traditional industries that have supported manufacturing have remained stuck in a structural depression. Kyoto has been losing its vigor with the slowdown of manufacturing.

Under such circumstances, the Kyoto Association of Corporate Executives and the Kyoto Chamber of Commerce and Industry have striven to develop new venture businesses in cooperation with top managers of the Kyoto high-tech companies.

Masao Horiba has taken the lead in these efforts, serving as chairperson of a special committee for the invigoration of Kyoto venture businesses at the Kyoto Association of Corporate Executives. In cooperation with local governments, Horiba has managed research parks, incubators, venture capital funds, and other organizations and facilities to support venture businesses in Kyoto.

Kazuo Inamori established the "Seiwa Jyuku" school where business managers and entrepreneurs, not only in Kyoto but also in the rest of Japan, learn about business management philosophy. The school originated from Inamori's meeting with young business managers in Kyoto around 1983 and has developed into the present Seiwa Jyuku, a nationwide organization. Business managers and entrepreneurs are learning about Kyocera's ameba management and other management systems as well as Inamori's business management philosophy.

In the ways explained above, the Kyoto high-tech companies have engaged in their respective philanthropic activities to thank the local community and to contribute to society.

Chapter 8

Conclusion

Global Companies Born in the Ancient Capital of Kyoto

We have reviewed the "Kyoto Model" of business management emerging from the birth and development of Kyocera, Omron, Murata, Rohm, and Horiba as leading Kyoto high-tech companies.

There are many high-tech companies in Kyoto that are not covered in this book. They include Nidec Corp., which commands a dominant share of the global market for spindle motors, indispensable for hard disk drives, as well as Nintendo Co., which competes fiercely with Sony Corp. and Sega Corp. in the global market for home video game machines.

The five Kyoto high-tech companies' business management systems are so unique and varied that it is difficult to characterize them as "the Kyoto Model." Nevertheless, we have adhered to the title "the Kyoto Model" in a bid to emphasize that global high-tech companies have developed in the internationally famed ancient capital of Kyoto. There are global companies based on various locations in Japan, other than Tokyo and Osaka.

Before fiscal 1998, when even the Kyoto high-tech companies had difficulties increasing sales amid the deepening recession, these companies continued to expand sales even amid the postbubble recession until fiscal 1997. Their excellence becomes even clearer when their earnings indicators are compared with those for the electrical machinery industry as a whole.

The Kyoto high-tech companies have steadily focused their research and development on their respective core products, acquired top market shares, and grown into leading companies while developing products related to their core ones. They have taken advantage of leading market shares to erect barriers against rivals and obtain competitive advantage.

Kyoto high-tech company founders and their successors have demonstrated their strong entrepreneurial leadership and practiced business management based on their respective philosophies. These top managers have fully understood their respective companies' core competences, adopted product-oriented strategies and focused business resources on core products to differentiate their companies from competitors.

In organizational management, the Kyoto high-tech companies have minimized the size of management and control units to emphasize flexibility, maintaining the group dynamics particular to the Japanese style of business management. On the other hand, they have promoted in-house decentralization and allowed employees to become more conscious of participation in business management. This has boosted employees' morale and encouraged them to be innovative.

The Kyoto high-tech companies' cost control is very strict. They have steadily demonstrated that cost savings can bring about profits. This might have led to their high profits.

A common feature among leading Japanese manufacturers is their excellence in production technologies. The Kyoto high-tech companies have also exploited their excellent production technologies to obtain their competitive advantage. Continual improvement of production technologies is essential to their maintenance of top market shares. They started up as R&D-oriented ventures and have given top priority to R&D. These companies have carefully devised systems to quickly produce market-oriented products in the fast-changing business environment.

The Kyoto high-tech companies have used aggressive mergers and acquisitions to enhance their respective core competences. They have made strategic investments, acquiring other companies both at home and abroad. Such management styles prompted these companies to realize the importance of stock markets earlier than many other Japanese firms and positively promoted investor relations activities. This, together with their robust earnings, is responsible for their high stock prices.

Positive social contributions are also a major feature of the Kyoto Model. Based on their respective philosophies, the founders of the Kyoto high-tech companies have contributed profits earned through business operations to the local community in various ways, out of gratitude for the support they have received from the community.

Amid the present globalization of the world economy, information technology development has gradually reduced barriers to the movement of business resources including people, goods, money, and information. In Japan as well as the rest of the world, geographical constraints on business operations have declined.

Many people may suspect that the companies that are active in the ancient Japanese capital of Kyoto have adopted the typical Japanese style of business management. In fact, however, the Kyoto high-tech companies' founders, who pursued business opportunities in the U.S. market in the early days of their firms' histories and directly learned the essence of U.S.-style business management, have flexibly incorporated the essential components of U.S.-style management into their companies' management systems. As a result, their management systems have many components close to the U.S. style.

The Kyoto Model was not completed in one day. Based on their philosophies, the founders of the Kyoto high-tech companies developed their management systems through repeated trial and error. These companies started up with the Japanese style of business management and have developed their management systems into those close to the U.S. style while pursuing business efficiency and higher speed.

They introduced U.S.-style management measures, such as overseas public stock offerings and aggressive exploitation of mergers and acquisitions, naturally in the course of their globalization in their early years. They have not necessarily imitated the U.S. style to follow a trend. Their management systems represent the collection of knowledge they have obtained through their various survival efforts.

In contrast, big Tokyo-based companies may have failed to have opportunities to learn lessons from their overseas experiences, including success and failure, because they entered European and U.S. markets only after the Japanese market matured. Even if there were opportunities, their top managers might have failed to flexibly introduce U.S. management measures. As a result, they might have retained the old-style Japanese management systems.

Historically, the influence of the feudal system on Tokyo has been stronger than on Kyoto. So, the authoritarian way of thinking might have been more deeply rooted in Tokyo. Contrary to general expectations, we suspect that people in Tokyo care too much about their appearance and are less flexible in some senses.

Such differences are essential to business management and are reflected in the attitudes of top managers. The Kyoto high-tech companies have been desperate, in a good sense, over the whole range of business management from development of technologies to quality and cost control. Their top managers have not cared too much about their appearance in business management. They have tried to make real gains rather than nominal ones. This may be the reason why they have been able to flexibly learn lessons from their experiences, including failures, in the United States, and incorporate them into their management systems.

Pressure of Slowing Economic Growth

We do not necessarily believe that Japanese companies should introduce the whole of the U.S. style of business management. But we believe that it is natural for companies to have management systems that reflect their countries' respective history and culture.

If foreign systems have points that Japanese companies should learn, however, they may have to positively introduce such points. We have refrained from covering business management systems in the European Union. But Japanese companies should introduce the excellent components of the EU systems as well as the U.S. systems.

Japanese companies' very successful penetration into overseas markets was the reason why Japanese-style business management became subject to research globally in the 1980s. The Japanese economy grew fast on the development of heavy and chemical industries, including shipbuilders, between 1955 and 1974 and on that of automobile and electronics industries between 1975 and 1993. Especially, Japanese consumer electronics products including TVs and VCRs were very successful, dominating world markets with their excellent quality.

The confidence that was built by Japanese-style management thanks to the great success of Japanese companies has, however, been gradually lost as their competitiveness has declined since the burst of the economic bubble.

In fact, the organized knowledge-developing process, which made Japanese companies successful, has a negative side. The process leads to the duplication of knowledge within an organization, which affects business and other efficiency. The long-term growth of companies had been a tacit precondition for the success of such a process.

As long as companies continue to grow, any inefficiency of the organized knowledge-developing process can be covered by the growth. As growth slows down, however, companies lose anything to cover up inefficiency and such a process becomes burdensome.

Learning from U.S.-Style Business Management Again

The United States in the 1980s was in the same position as Japan is in now. In the 1980s, U.S. manufacturers of automobiles, precision machines, and consumer electronics had lost international competitiveness and had to undergo considerable restructuring. Then, U.S. manufacturers thoroughly studied Japanese-style business management represented by the just-in-time system that was all the rage at the time. They re-engineered their organizations to give priority to horizontal communications.

U.S. manufacturers strove to overcome their problems. They introduced fast-developing information technologies into their management systems to speed up decision-making and formed strategic alliances to make up for any gaps in their business resources. They thus reformed their management systems into flexible ones that could meet fast-changing customer needs promptly. This allowed them to recover.

Currently, Japanese companies are under pressure to urgently rebuild their management systems to survive the great business environment changes including the globalization of economic activities, the protracted stagnation of the Japanese and Asian economies, and the dollar's long-term downward trend against the yen.

Just as Japanese business managers did to rebuild the Japanese economy just after the war, present-day managers have to urgently learn U.S.-style business management again and introduce the best components of the U.S. style. In order to reinvigorate Japanese companies and the economy, Japanese managers should reform their management systems while keeping the strong points of the Japanese style that has been developed so far. We are confident that Japanese business managers have the capability to do so.

Japanese companies can incorporate the strengths of U.S.-style business management into the Japanese style to improve their management systems. For Japanese companies currently under pressure to meet global standards, the challenge is how to integrate the two styles of business management. Some hints for their intelligent integration may be found in the Kyoto Model.

Financial Data of the Kyoto High-Tech Companies (1988–2004)

		Kyocera	
	Sales	Operational Profit	Pretax Profit
1988	300,409	41,563	50,709
1989	338,704	54,277	60,124
1990	421,032	60,026	68,235
1991	461,233	55,580	60,063
1992	453,499	42,298	54,555
1993	431,599	39,374	46,188
1994	427,698	43,503	41,486
1995	498,566	66,358	64,668
1996	647,155	118,455	127,873
1997	714,765	124,801	131,399
1998	725,312	95,500	105,380
1999	725,300	55,800	61,800
2000	812,600	92,200	97,500
2001	1,285,100	207,200	400,200
2002	1,034,600	51,600	55,400
2003	1,069,800	83,400	76,000
2004	1,140,800	109,000	115,000

Omron

	Sales	Operational Profit	Pretax Profit
1988	315,618	28,689	25,770
1989	372,447	45,009	39,670
1990	416,231	45,347	43,701
1991	464,376	45,678	43,429
1992	483,247	26,608	19,670
1993	462,702	20,689	10,042
1994	460,869	19,730	13,055
1995	489,700	33,478	24,948
1996	525,289	39,239	32,252
1997	594,261	40,905	39,248
1998	611,795	46,032	42,243
1999	555,280	11,849	8,249
2000	555,358	26,180	21,036
2001	594,259	44,349	40,037
2002	533,964	4,221	(25,373)
2003	535,073	32,313	4,732
2004	584,889	51,403	47,984

Murata

	Sales	Operational Profit	Pretax Profit
1988	211,695	38,276	42,548
1989	243,092	43,376	48,590
1990	247,776	35,691	41,687
1991	277,367	44,894	51,552
1992	280,467	44,578	51,525
1993	272,054	40,083	45,402
1994	279,234	46,712	51,063

Murata

	Sales	Operational Profit	Pretax Profit
1995	317,451	71,189	73,618
1996	321,859	69,629	73,668
1997	330,612	61,589	65,405
1998	362,252		72,694
1999	367,048	57,061	61,627
2000	459,125	100,767	108,074
2001	584,011	174,248	173,925
2002	394,775	51,001	52,408
2003	394,955	59,187	59,094
2004	414,247	74,210	78,685

Rohm

	Sales	Operational Profit	Pretax Profit
1988	117,570	7,416	6,629
1989	146,117	12,641	12,991
1990	166,882	11,113	13,552
1991	189,476	11,355	11,509
1992	204,937	22,856	19,081
1993	186,897	16,556	15,132
1994	199,988	24,143	21,557
1995	241,492	43,286	41,465
1996	292,280	72,569	71,579
1997	297,789	77,738	82,914
1998	335,922	116,502	110,064
1999	328,631	90,091	95,284
2000	360,079	122,342	122,581

	Rohm		
	Sales	Operational Profit	Pretax Profit
2001	409,335	137,743	148,136
2002	321,264	66,458	75,041
2003	350,281	96,122	91,684
2004	355,630	94,507	92,083

	Horiba		
	Sales	Operational Profit	Pretax Profit
1988	20,939	1,776	2,080
1989	23,095	2,214	2,580
1990	25,964	2,487	3,019
1991	30,581	2,854	3,509
1992	35,385	3,081	3,629
1993	43,970	2,641	2,679
1994	40,812	2,335	1,898
1995	38,287	1,944	1,872
1996	40,674	2,694	2,763
1997	50,315	3,692	3,346
1998	62,425		5,464
1999	67,597	2,914	2,775
2000	71,030	3,817	3,498
2001	77,872	4,749	4,798
2002	74,467	2,547	1,599
2003	78,501	5,473	3,766
2004	85,072	6,850	5,597

Consolidated Sales

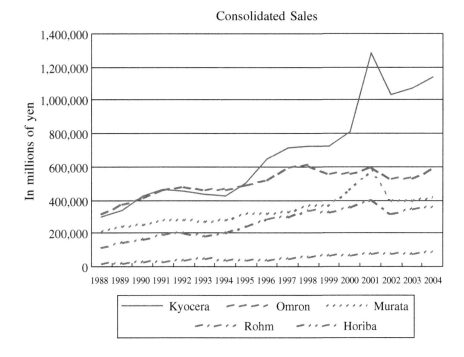

Consolidated Pretax Profit

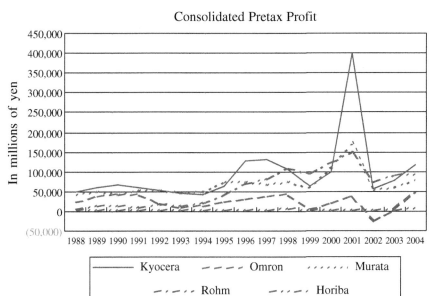

Supplementary Remarks

During the recession of the 1990s many Japanese companies collapsed. However, the sales and profit of Kyoto high-tech companies grew consistently even under the recession, as shown in the above figures.

Since then, the Japanese economy has been growing slowly. Even under the difficult situation where companies are having difficulty expanding their sales and profit, Kyoto high-tech companies have not changed their management style. They rather concentrate their corporate resources on their core competence.

Kyoto high-tech companies continue to expand their global network, especially in China and the Southeast Asia region. They establish subsidiaries in many countries, and introduce their own unique management systems to these subsidiaries.

At the same time, Kyoto high-tech companies also try to match their management systems with regional customs. They make use of their unique small size and effective organization structure and are able to strengthen their core products through improvement of production control. They emphasize the same management philosophies to implement corporate strategy in the subsidiaries. Kyoto high-tech companies thus are able to succeed on a sustainable basis.

Kyoto high-tech companies adopt a positive attitude towards environmental protection activities and social contribution, although their businesses are going through a rough patch. This affected their stock's value as socially responsible investment (SRI) diffused in the late 1990s, although their sales and profits are generally increasing on a steady basis.

The management system of Kyoto high-tech companies can be said to be one of the best corporate management models even in the 21st century.

Bibliography

Books (Written in Japanese)

Abe, Kazuyoshi. *A Study on Kazuma Tateisi.* Kano Shobo, 1991.

Iida, Fumihiko. *Key Points of Japanese-Style Business Management.* PHP Institute Inc., 1998.

Itami, Takayuki. *Three Waves for Japanese Industry.* NTT Publishing Co., 1998.

Inamori, Kazuo. *Respect the Divine and Love People: What Have Supported My Management.* PHP Institute Inc.

Inamori, Kazuo. *Kazuo Inamori's Practical Theory: Management and Accounting.* Nihon Keizai Shimbun Inc.

Imai, Kenichi; Komiya, Ryutaro, ed. *Japanese Companies.* University of Tokyo Press, 1989.

Usui, Shinichi; Diamond Harvard Business Desk, ed. *Creating a 21st Century Company: Omron's 21st Century Vision Golden 1990s.* Diamond Inc., 1991.

Ogawa, Tadao. *Omron Realizing Dreams.* Japan Management Association Management Center, 1997.

Okumura, Akihiro. *Japan's Top Management.* Diamond Inc., 1982.

Odagiri, Hiroyuki. *Japanese Companies' Strategy and Organization.* Toyo Keizai Inc., 1992.

Kagono, Tadao; Nonaka, Ikujiro; Sakakibara, Kiyonori; Okumura, Akihiro. *Comparison between Japanese and U.S. Business Management Styles.* Nihon Keizai Shimbun, Inc., 1983.

Kagono, Tadao. *Recovery of Japanese-Style Management.* PHP Institute, Inc., 1997.

Kunitomo, Ryuichi. *Kyocera's Kazuo Inamori: Vigorous and Forethoughtful Management.* Pal Publishing Inc., 1996.

Kunitomo, Ryuichi. *Kyocera's Amoeba Management System: Retractable System to Save Cost and Pursue Higher Efficiency.* Pal Publishing Inc., 1997.

Shiba, Ryotaro. *Hand-Cut Japanese History.* Bunshun Co., 1990.

Shiba, Ryotaro. *Shape of This Country II.* Bunshun Co., 1993.

Nakazawa, Takao. *New Era for Small Companies.* Iwanami Shoten, 1998.

Nikkei Business, ed. *Why is This Company Strong?: Road to Visionary Company.* Nikkei Business Publications, Inc., 1996.

Nikkei Business, ed. *Bullish Company 3M.* Nikkei Business Publications, Inc., 1998.

Nihon Keizai Shimbun, Inc., ed. *Market Shares '98.* Nihon Keizai Shimbun, Inc., 1997.

Nonaka, Ikujiro; Takeuchi, Hirotaka; Umemoto, Katsuhiro. *Knowledge-Developing Companies.* Toyo Keizai Inc., 1996.

Hisada, Soya; Nishimura, Hiromichi, ed. *Hidden Stories of Kyoto: History and People.* Dohosha Publishing Co., 1997.

Horiba, Masao. *Go Away If You Are Dissatisfied: New Relations between Employees and Companies.* Nihon Keizai Shimbun, Inc., 1995.
Horiba, Masao. *Masao Horiba's Management Memos.* Toyo Keizai Inc., 1998.

Books (Translated into Japanese)

Hammel, G.; Prahalad, C.K. *Core Competence Management* (translation of the original *The Core Competence of the Corporation* in English). Nihon Keizai Shimbun, Inc., 1995.
Jasinowski, Jerry; Hamrin, Robert. *Revival of American Manufacturers: Proven Paths to Success form 50 Top Companies* (translation of the original *Making It in America: Proven Paths to Success from 50 Top Companies*). Tokyu Agency Inc., 1996.
Galbraith, Jay R.; Nathanson, Daniel A. *Management Strategy and Organization Design* (translation of the original *Strategy Implementation: The Role of Structure and Process*). Hakuto Shobo, 1989.
Hammer, Michael; Champy, James. *Reengineering Revolution* (translation of the original *Reengineering The Corporation: A Manifesto for Business Revolution*). Nihon Keizai Shimbun, Inc., 1993.
Porter, Michael. *Competitive Advantage Strategy* (translation of the original *Competitive Advantage*). Diamond Inc., 1985.
Porter, Michael. *Competitive Strategy* (translation of the original *Competitive Strategy*). Diamond Inc., 1995.
Polanyi, Michael. *Tacit Knowledge Dimension* (translation of the original *The Tacit Dimension*). Kinokuniya Co., 1980.

Journal Articles (in Japanese)

Japan Company Handbook. Summer 1998, Toyo Keizai Inc.
Weekly Toyo Keizai Extra Issue: Economic Statistics Almanac. Toyo Keizai Inc., 1998.
Weekly Toyo Keizai Extra Issue: Corporate Financial Carte. Toyo Keizai Inc., 1998.
Nikkei Annual Consolidated Corporate Reports. Nihon Keizai Shimbun, Inc., 1998.
Japan Market Share Handbook. Yano Research Institute, 1998.
"Why Were Interesting Companies Born in Kyoto?" *Will.* April 1991, pp. 86–89.
"Kyocera-Yashica Merger (1)." *Accounting Information.* May 20, 1994, pp. 12–18.
"Horiba's Time-One-Half Movement Featuring No Overtime Work Day and No Telephone Time." *Monthly Personnel Managment.* October 1995, pp. 112–115.
"Structural Reforms That Achieved Four-Year Consecutive Growth in Sales and Profit after Restructuring," *Zaikai.* January 22, 1997, pp. 39–41.
"Kyocera Exploits New M&A Method to Get Overseas Cornerstone." *Weekly Diamond.* February 24, 1990, pp. 114–116.
"Experiences of Refreshment System-Initiating Companies." *Weekly Diamond.* May 18, 1991. pp. 90–93.
"A Study on Depression-Resistant Kyoto Companies." *Weekly Diamond.* January 30, 1993, pp. 70–75.
"Unique Management Approach to Hit R&D Targets." *Weekly Diamond.* June 8, 1996, pp. 50–53.
"Scorecards for Japanese Firms' Overseas M&A Deals." *Weekly Diamond.* July 6, 1996, pp. 88–91.

"Leading IC Manufacturer Developed with Founder's Will and Endeavor." Weekly Toyo Keizai. November 9, 1984, pp. 60–61.

"How Will Rohm Use Ultimate Memory? — Minor Firm's Global Domination Operation." Weekly Toyo Keizai. March 25, 1995, pp. 74–77.

"Becoming A Venture Again: In-House Deregulation for a New Offensive." Weekly Toyo Keizai. April 13, 1996, pp. 124–127.

"Psychology Management to Produce Surprising Profit." Weekly Toyo Keizai. July 5, 1997, pp. 152–155.

"We Will Continue to Be Venture." Weekly Toyo Keizai. October 11, 1997, pp. 92–95.

"Rohm: Rules of Gambler Continuing to Win." Weekly Toyo Keizai. March 11, 1998, pp. 132–136.

"Top Global Companies in Japan." The 21. May 1998, pp. 81–93.

"Horiba's Field-Based Manager System and Merit-Based Personnel Evaluation." Wage Practices. March 1, 1994, pp. 24–31.

"System to Achieve Speedy Management: Producing only Necessary Goods." Nikkei Computer. October 14, 1996, pp. 130–135.

"Focusing on Authority Transfer and Information Sharing: Strict Follow-up on Collective Decision-Making." Nikkei Information Strategy. December 1996, pp. 148–150.

"Eying Market Left by Big Companies: Operation Third Player." Nikkei Business. February 27, 1989, pp. 68–72.

"Venture Spirit Supports Pursuit of World's No. 1 Position." Nikkei Business. April 27, 1992, pp. 88–89.

"Uniqueness and Boldness Eliminating Follow-the-Leader Mentality: Better Performance Than Big Companies' in Semiconductor Business." Nikkei Business. March 8, 1993, pp. 90–93.

"Information Collection Is Key to Profitability Improvement: Waiting for Chance with Surplus Personnel." Nikkei Business. November 22, 1993, pp. 60–63.

"Original Processes as Well as Original Products: Winning with Black Box Strategy." Nikkei Business. June 12, 1995, pp. 20–41.

"Visionary Management: Spearhead Companies Have Philosophies." Nikkei Business. August 21, 1995, pp. 20–41.

"Young Power for New Businesses: Leading to Institutional Deregulation." Nikkei Business. November 25, 1996, pp. 42–44.

"Logics Underlie Capitalism: Good Practices Producing Profit." Nikkei Business. December 30, 1996, pp. 120–124.

"ROE Is Not Almighty: Reconsidering Measures of Companies." Nikkei Business. June 2, 1997, pp. 30–31.

"Murata Takes Advantage of Crisis: \1-for-One-Unit Approach for Excellent Profitability." Nikkei Business. May 11, 1998, pp. 22–32.

"Promotion of Business Expansion for New Century: Technology-Oriented Omron." Forbes (Japanese version). January 1997, pp. 94–99.

"Rohm: An Excellent Company for 21st Century." Foresight. April 1998, pp. 78–79.

Index

ABX S.A., 54
acid rain, 99
Akira Kurosawa, 97
ameba, 68–71, 88
ameba management, 66, 68–71, 88, 101
automobile emission analyzers, 44, 46, 48, 99
AVX, 40, 47, 50–52

big-company disease, 73
black-box technologies, 61
bubble economy, 33
bureaucratic dynamics, 5–7, 9

core competence, 18, 19, 32, 50, 53, 66, 78, 91, 104
craftsmen, 31, 32
cross-shareholding, 4
custom IC, 37, 43, 46, 84, 86
Cybernet, 40, 50, 51

DDI Cellular Group, 46
decentralization, 9, 22, 65, 66, 71, 73, 74, 104
disclosure, 95, 96

economic bubble, 13, 15, 107
Elco, 40, 50, 51
electronic collective decision-making systems, 75, 78
Exar, 43, 53
explicit knowledge, 3

family business, 33
field-based manager system, 90–92

General Electric, 42, 48
golden products, 85
golden technology system, 85
group dynamics, 5–9, 22, 65, 83, 104

Hewlett-Packard, 8, 11, 66
hierarchic organization, 5
Horiba, 11, 13, 19, 22, 25, 30, 31, 34, 37–39, 44–50, 53, 54, 57, 62, 80, 81, 83, 85, 89–92, 94–96, 99, 100, 103
horizontal organization, 78
hourly profitability, 22, 69, 70, 88
hourly value added, 70, 88

in-house awards, 22
in-house magazine, 94
in-house manufacturing, 25
in-house manufacturing of production, 22, 87
in-house materials development, 87
in-house prize, 44, 89, 90
in-house technology database, 85
in-house unions, 1
incubator, 100
innovation, 3, 8, 9, 18–27, 68, 80, 83, 89, 90

Japanese-style management, 1, 2, 7, 8, 27, 107, 108
job rotation, 2
just-in-time system, 2, 107

kanban, 2
knowledge development, 3, 8, 18, 19

Kyocera, 8, 11, 13, 15, 19, 22, 25, 34,
  37–40, 45–47, 49–52, 56–59, 63, 66,
  68–72, 84, 88, 95, 96, 99, 101, 103
Kyoto International Music Students
  Festival, 97
Kyoto Prize, 96, 97
Kyoto University, 32, 90
Kyotoism, 31, 32

lifetime employment, 1, 2

3M, 8, 11, 66
main bank, 4
market-oriented products, 22, 80, 84, 85,
  104
matrix management, 22, 66, 76–78, 88, 89
matrix organization, 66, 81
matrixes, 77
Meiji Period, 31, 33
Meiji restoration, 33
meisters, 22, 90–92
merger and acquisitions (M&A), 40, 104,
  105
merit-based system, 91, 92
modular products, 61, 87, 88
morale, 18, 19, 21, 22, 34, 52, 56, 59, 71,
  84, 88, 91–93, 104
Motorola, 42, 48
Murata, 11, 13, 15, 19, 22, 25, 34, 37–39,
  41, 42, 46–50, 53, 57, 59, 61, 64, 66,
  76–78, 83–89, 92–95, 97, 99, 103

network organization, 66, 68
new business development centers, 22, 85,
  93
New York Stock Exchange, 47, 49, 52
Nidec Corp., 103
Nintendo, 103
Nippon Data, 50, 52
Nonaka Ikujiro, 3

Omron, 11, 13, 19, 25, 34, 37–41, 46–50,
  52, 56, 57, 59, 60, 63, 66, 71, 73–75,
  83, 85, 89, 90, 93–95, 98, 99, 103
Omron Taiyo, 98
operation-oriented strategy, 5, 6, 9

per-hour earnings, 69
per-hour value added, 69
philanthropy, 25, 26, 96, 98
philosophy, 19, 21, 55–64, 87, 92, 96–101
process cost control, 22
producer system, 66, 71, 73, 74
product-oriented management, 22, 80
product-oriented management system, 22,
  80
product-oriented strategy, 5–8, 83, 104
production technology, 1, 25, 26, 32, 33,
  79, 86, 87, 104

quality control, 2

recycling, 98, 100
return on asset (ROA), 15, 74
return on equity (ROE), 13, 95, 96
Rohm, 11, 13, 15, 19, 22, 25, 34, 37–39,
  42, 43, 46–50, 53, 57, 61, 78, 79, 84, 86,
  87, 89, 90, 92, 94, 95, 97, 99, 100, 103
Rohm Music Foundation, 25, 97
Ryotaro Shiba, 31

San Francisco Stock Exchange, 39, 49
Seiwa Jyuku, 100, 101
self-supporting units, 22, 69
seniority-based salaries and promotions, 1, 2
Silicon Valley, 39, 43, 48, 53
Singapore Stock Exchange, 39, 49
social needs theory, 41, 73, 75
social welfare, 25, 98
Sony, 42, 51, 98, 103
strategic development process management
  system, 84, 85
sustainable business management, 32, 33

tacit knowledge, 5
Tadao Kagono, 1, 5, 27
TDK, 8, 11
Texas Instruments, 39, 47
top-down, 4, 5

venture capital funds, 100
venture capital system, 85, 93

Yashica, 40, 50, 51

Printed in the United States
By Bookmasters